THE OCEANS

WARWICK PRESS · NEW YORK

Editorial

Author
David Lambert

Editorial Consultant
Tony Loftas

Editor
Angela Wilkinson

Assistant Editor
Christine King

Dictionary Compiled by
Tony Loftas

Editor's note: On pages 84 to 91 of this book you will find a dictionary of oceans. Many of the terms used throughout the main chapters are printed in SMALL CAPITALS. You will find an entry in the dictionary which will give you an explanation of these terms.

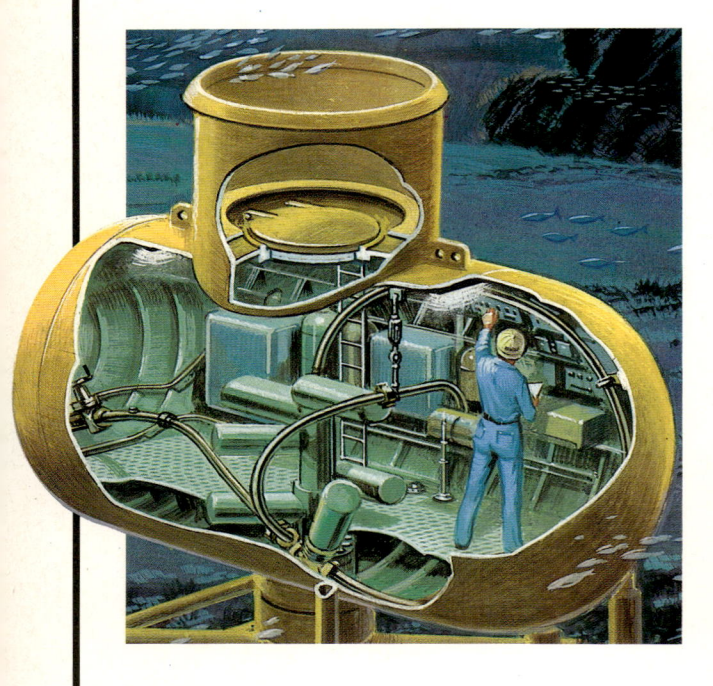

Published 1980 by Warwick Press,
730 Fifth Avenue, New York, New York 10019.

First published in Great Britain by
Ward Lock Ltd., London, in 1979.

Copyright © 1979 by Grisewood & Dempsey Ltd.

Printed in Italy by Vallardi Industrie Grafiche, Milan.

6 5 4 3 2 1   All rights reserved

Library of Congress Catalog Card No. 79-93266

ISBN 0-531-09168-6

Contents

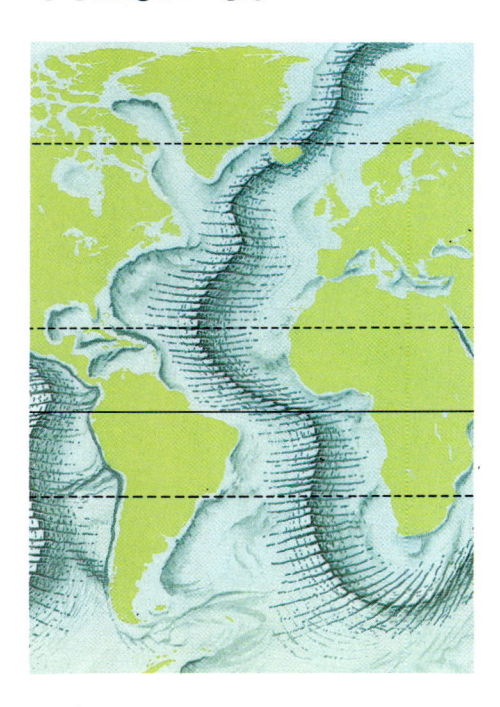

Chapter One
THE OCEANS 8
Sizing Up the Oceans 10
How the Sea-Floor Is Made and Lost 12
Submarine Landscape 14
Looking at the Oceans 16

Chapter Two
THE OCEAN SURFACE 18
Ships 20
Currents 22
Waves 24
Wave Energy 26
Tides 27
Changing Sea Levels 28
Ice at Sea 30

Chapter Three
LIFE IN THE WATERS 32
The Flow of Food 34
The Undersea World 36
Sea Mammals 38
Life in the Depths 40
Polar Regions 41
Tropical Seas 42
Fishing 44

Chapter Four
MAN BENEATH THE SEA 46
Divers 48
Underwater Habitats 50
Working Under Water 52
Submarines and Submersibles 54

Chapter Five
WEALTH FROM THE OCEANS 56
Mining the Waters 58
Mining the Seabed 60
Oil and Gas 62
Sunken Treasure 64

Chapter Six
AT THE OCEAN'S EDGE 66
Sculpting the Shores 68
Life on the Shore 70
Seaweeds 72
Islands 74
Using the Shore 76

Chapter Seven
TOMORROW'S OCEANS 78
Endangered Oceans 80
Schemes for the Future 82

Dictionary of Oceans 84

Index 92

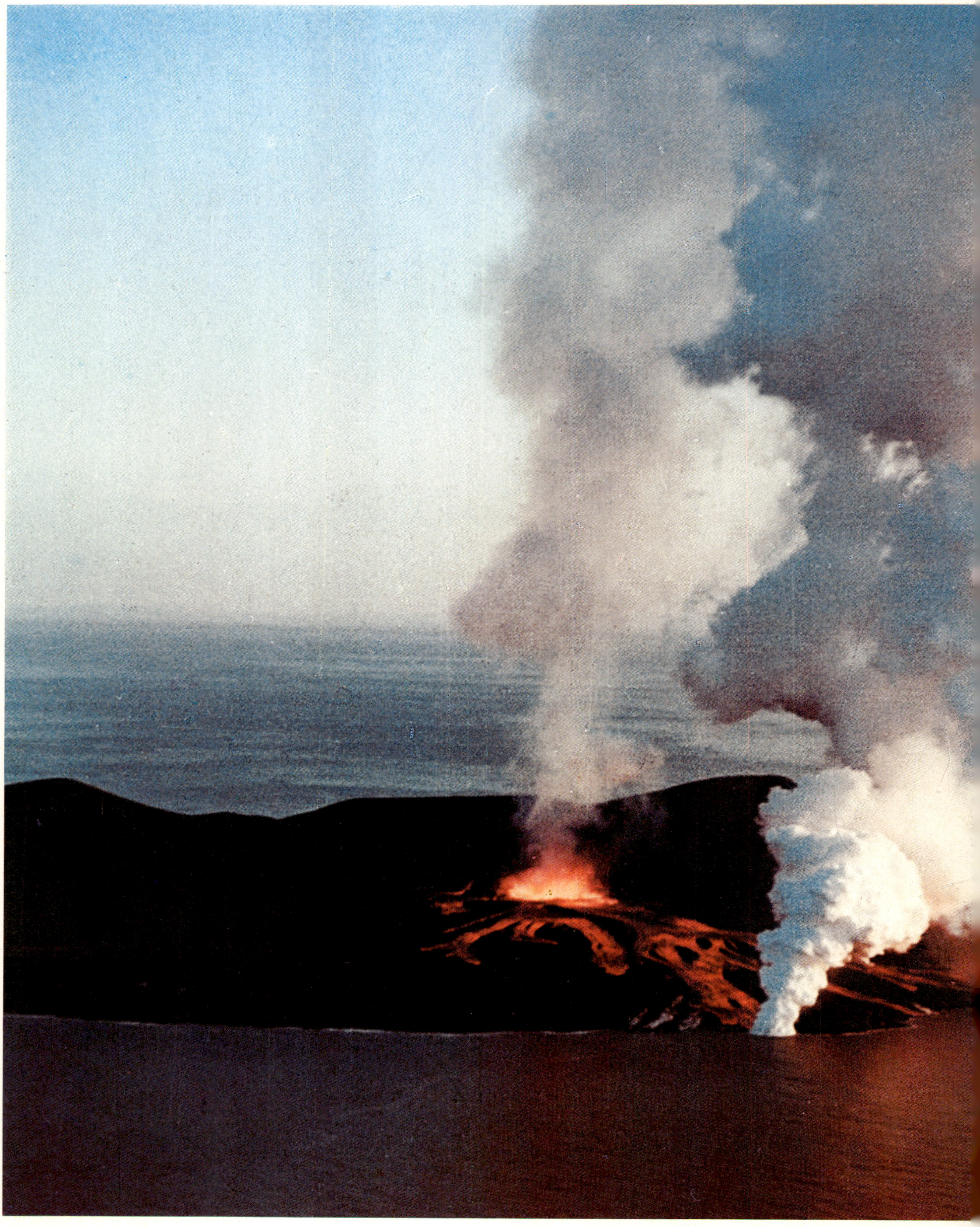

Chapter One

THE OCEANS

There were no oceans 4600 million years ago, when the Earth took shape, most likely from a vast shrinking cloud of dust. Crushed and heated by the shrinking process, its solids would have largely changed to molten rock. The lightest elements rose towards the surface. Some escaped as gases. Others formed a seething sea of liquid rock. In time, surface rocks cooled and hardened into rafts that coalesced to build a solid crust. Meanwhile, volcanoes spewed steam and other gases into the hot, prehistoric atmosphere.

After millions of years, the atmosphere cooled enough for the steam to change to water vapour that condensed to droplets large enough to fall as rain. What followed was a deluge lasting tens of centuries.

When it ended, vast tracts of basaltic rock were covered by rainwater: the primeval oceans had been born. Only rafts of granitic rock – less dense than basalt – rode high and dry to form the first continent.

The volcanic island of Surtsey erupting in the sea pours smoke and steam into the atmosphere. The very waters of the sea may come from steam produced by prehistoric volcanoes.

Sizing Up the Oceans

Roughly seven-tenths of the Earth's surface lies beneath a great tract of salty oceanic water. Continents separate this into four main, connected regions: the Arctic, Atlantic, Indian and Pacific oceans. Some geographers consider that the southern ends of the last three make up a fifth ocean – the Southern, or Antarctic, Ocean.

Of the four main oceanic regions, the Pacific, east of Asia and Australia and west of the Americas, is easily the largest – larger than all continents combined. It also has the greatest depths. The Arctic Ocean is the smallest, shallowest, and coldest ocean of them all.

Why Seas Are Salty

All the oceans lose water in water vapour drawn up by the Sun's heat. This vapour condenses in cool air, and forms clouds which may later release rain, sleet, hail or snow. More than a quarter of the water lost from oceans falls upon the land. From there, rivers bear it back down to the sea. This process is known as the WATER CYCLE: it keeps the oceans full and makes their waters salty. Minerals washed off the land travel to the sea suspended and dissolved in rivers. When seawater evaporates, these substances remain behind.

Sodium chloride ('salt') makes up more

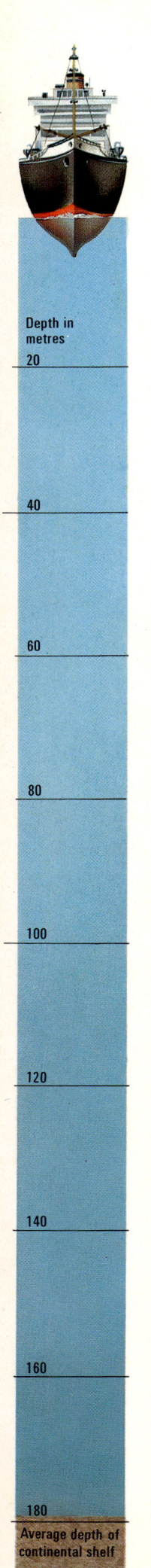

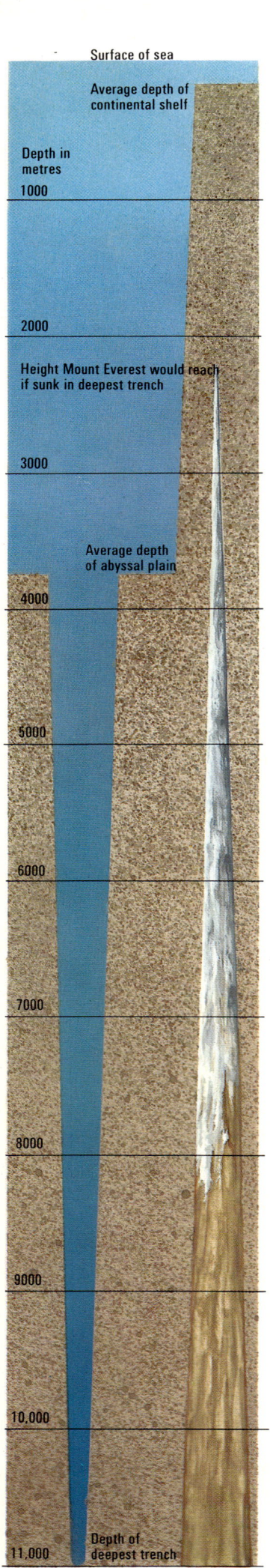

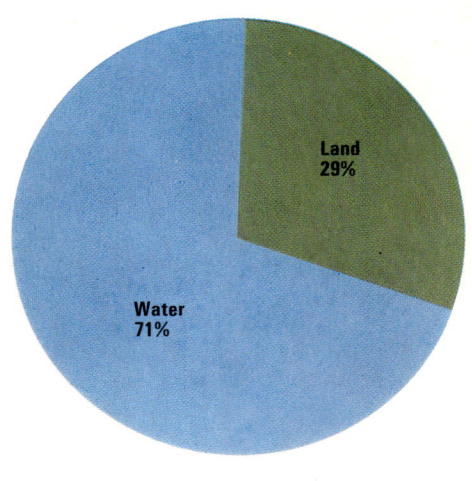

Above: The proportions of the Earth's surface which are land and water: about 71 per cent is water.

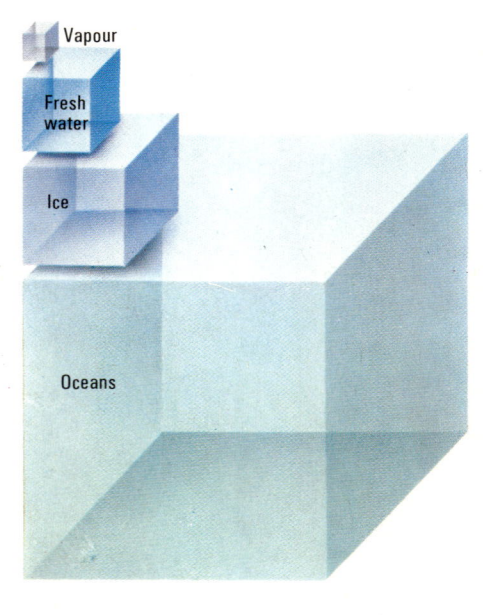

Above: Oceans hold 97 per cent of the world's water. More than 2 per cent is in ice; less than 1 per cent is in fresh water and air.

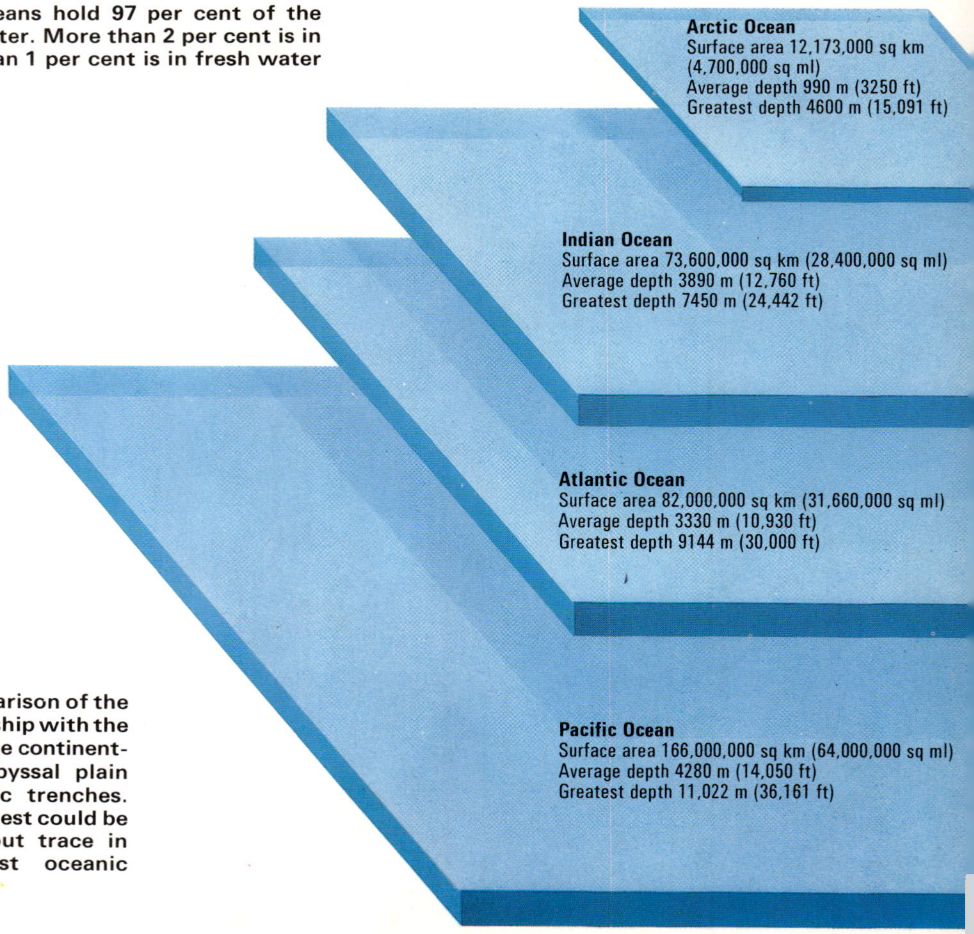

Left: Comparison of the height of a ship with the depths of the continental shelf, abyssal plain and oceanic trenches. Mount Everest could be sunk without trace in the deepest oceanic trench.

Arctic Ocean
Surface area 12,173,000 sq km (4,700,000 sq ml)
Average depth 990 m (3250 ft)
Greatest depth 4600 m (15,091 ft)

Indian Ocean
Surface area 73,600,000 sq km (28,400,000 sq ml)
Average depth 3890 m (12,760 ft)
Greatest depth 7450 m (24,442 ft)

Atlantic Ocean
Surface area 82,000,000 sq km (31,660,000 sq ml)
Average depth 3330 m (10,930 ft)
Greatest depth 9144 m (30,000 ft)

Pacific Ocean
Surface area 166,000,000 sq km (64,000,000 sq ml)
Average depth 4280 m (14,050 ft)
Greatest depth 11,022 m (36,161 ft)

than three-quarters of the dissolved salts in seawater, followed by magnesium chloride, sodium sulphate and many others. There are also scarce 'trace' elements, some vital to sea plants and animals.

Overall, salinity stays more or less constant, but varies from place to place. It is between 32 and 37 parts per thousand in the open ocean, but as little as 2 parts per thousand in areas of the cool, river-fed Baltic Sea, and up to 40 parts per thousand in the hot, dry-shored Red Sea.

Temperature and Pressure

Like salinity, temperature varies according to the depth of the water and its geographical location. Surface temperatures range from 29°C (85°F) in the Red Sea in summer to −2°C (28°F) – the temperature at which seawater freezes – in polar seas in winter. Because oceans are heated from above, temperature in the open oceans drops sharply at the THERMOCLINE, a boundary 30–80 metres (100–260 feet) down. *Average* temperature is only 3.8°C (39°F).

Local differences in temperature and salinity produce differences in density. These help to trigger deep-sea currents (page 23).

Lastly, pressure increases with depth, by 1 atmosphere per 10 metres: from 1 atmosphere – 1 kilogramme per square centimetre (14.7 pounds per square inch) – at the surface to nearly 1150 times greater at 10,917 metres (6.78 miles), the greatest depth attained by a manned vehicle. Water is normally incompressible, but in the deepest ocean trenches it is squashed under its own weight by as much as 30 metres (100 feet).

Below: The surface areas and average depths of the four main oceanic regions. The Pacific alone covers more than one-third of the globe.

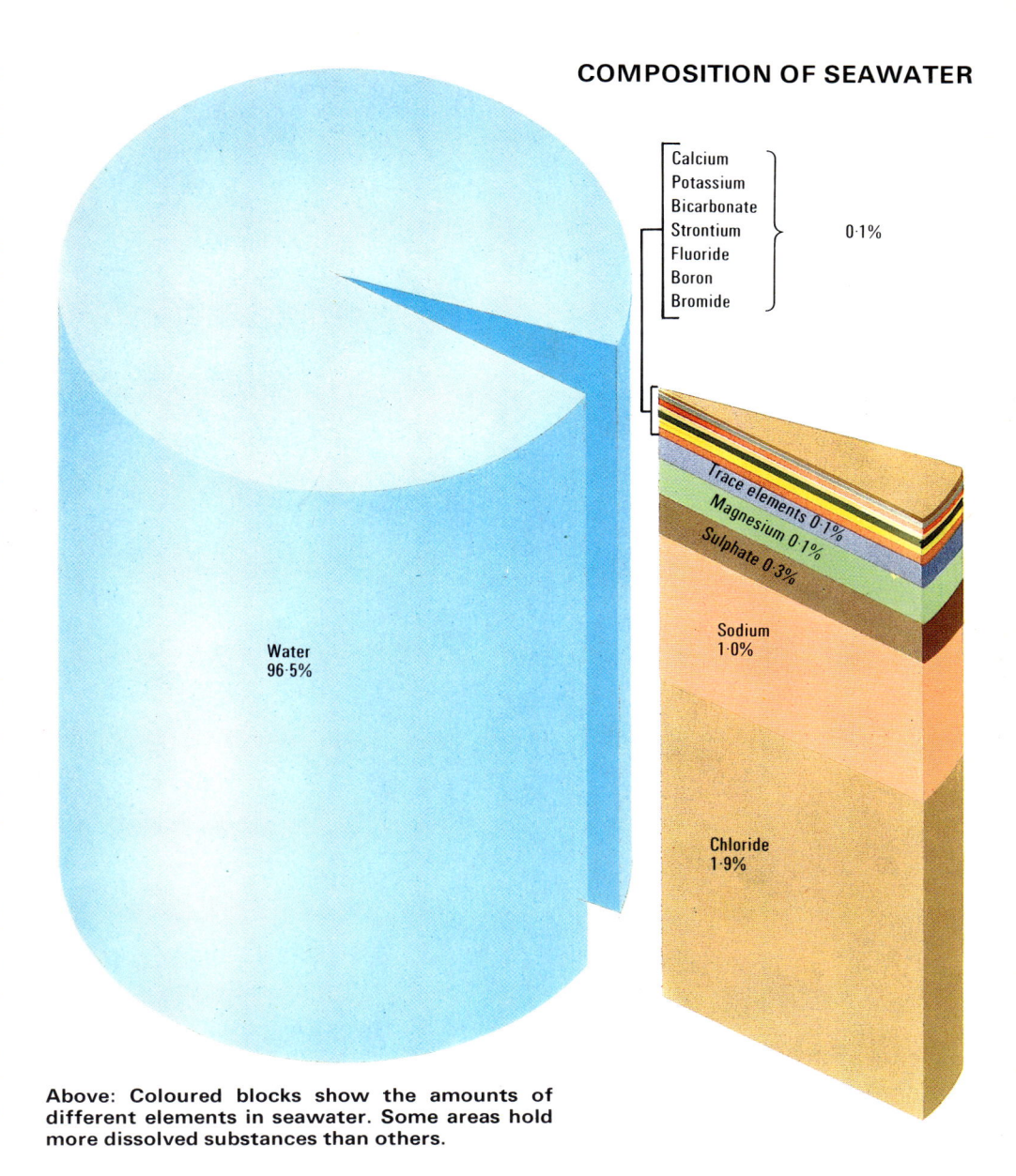

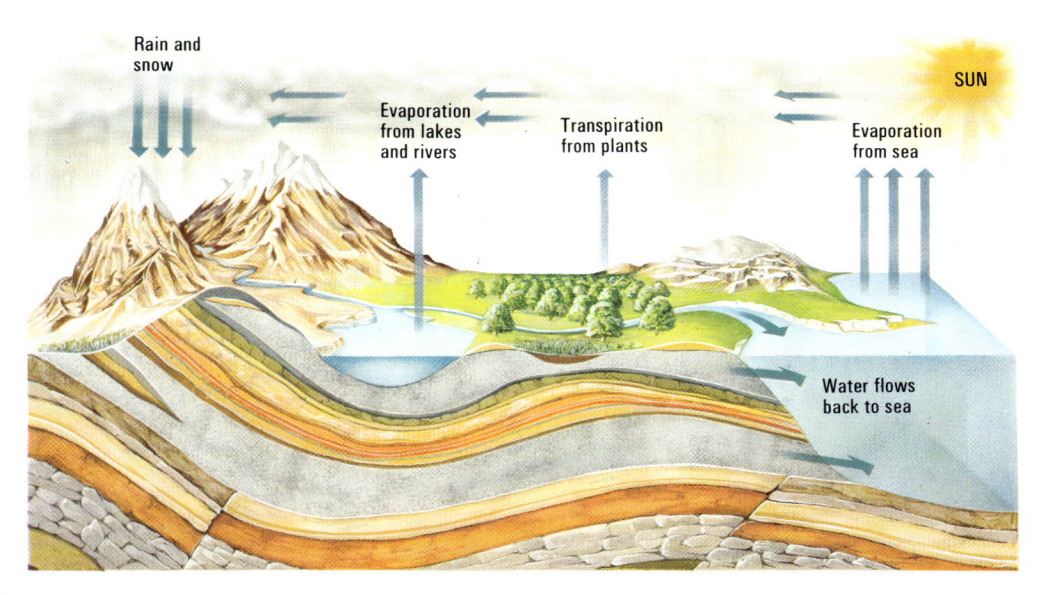

COMPOSITION OF SEAWATER

Above: Coloured blocks show the amounts of different elements in seawater. Some areas hold more dissolved substances than others.

Above: The water cycle starts as the Sun's heat sucks up water vapour from the surface of the ocean. Winds blow much of this water vapour over the land. The vapour cools as it rises or if it meets cold air. Cooling causes the vapour to condense into droplets that form clouds. If droplets join, they grow too big and heavy to remain in air, and fall as rain, sleet, hail or snow. Some of this precipitation forms ice; some feeds lakes and rivers; some soaks into rocks; some evaporates. But much of it flows back into the oceans.

THE DRIFTING CONTINENTS

Right: Maps show how the drifting continents have changed position in the last 200 million years. Once, all were joined in one super-continent known as Pangaea ('All Earth'). By 135 million years ago Pangaea had broken up into huge northern and southern continents, which in turn split into smaller pieces that began to drift apart. Arrows show which way they went. For instance, North America split off from Europe, and South America split off from Africa. Australia and Antarctica separated. But India crashed into Asia. These shifts produced the world map that we know today. But the lands are still in motion.

200 million years ago

PANGAEA

135 million years ago

Today

Below: Earth's crust consists of plates moving at 1–2 centimetres per year in different directions, shifting the continents and ocean basins. Where plates crash or separate, they unleash volcanoes and earthquakes.

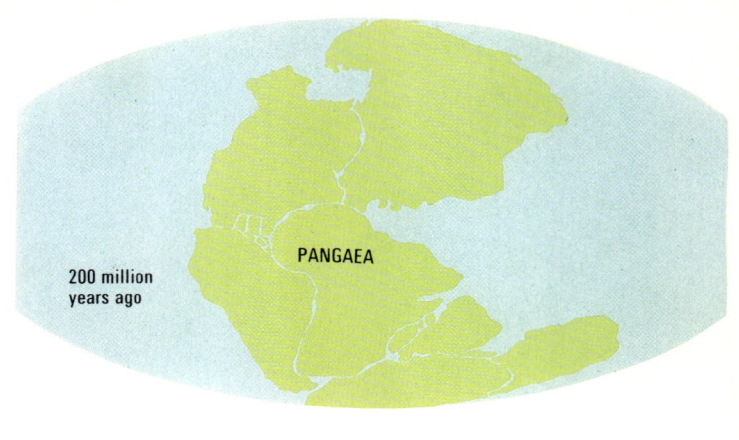

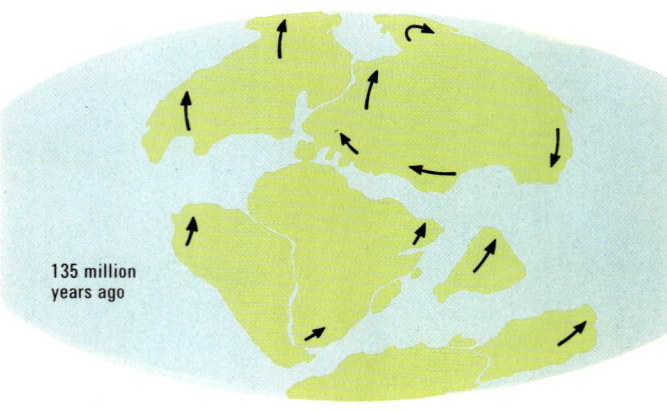

Above: A satellite-eye-view of the gulfs of Suez and Aqaba at the northern end of the Red Sea. This slender sea formed when Africa and Asia split apart 10 million years ago. If the rift widens, the Red Sea may open up into an ocean. The Atlantic formed in such a way.

Below: Where two crustal plates separate in mid ocean, molten rock wells up to form a spreading ridge. An oceanic plate colliding with a continental plate is driven down beneath it – a process called subduction.

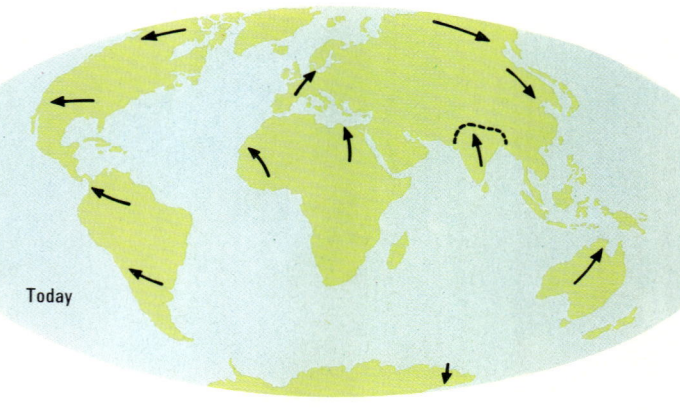

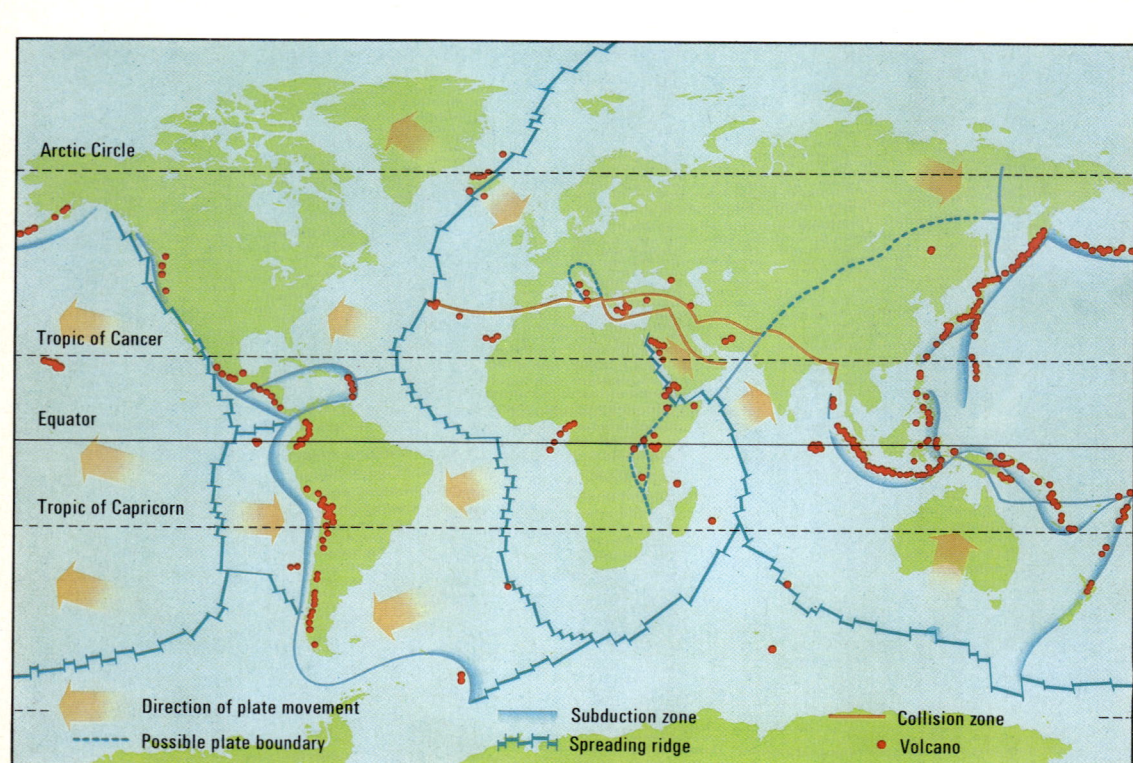

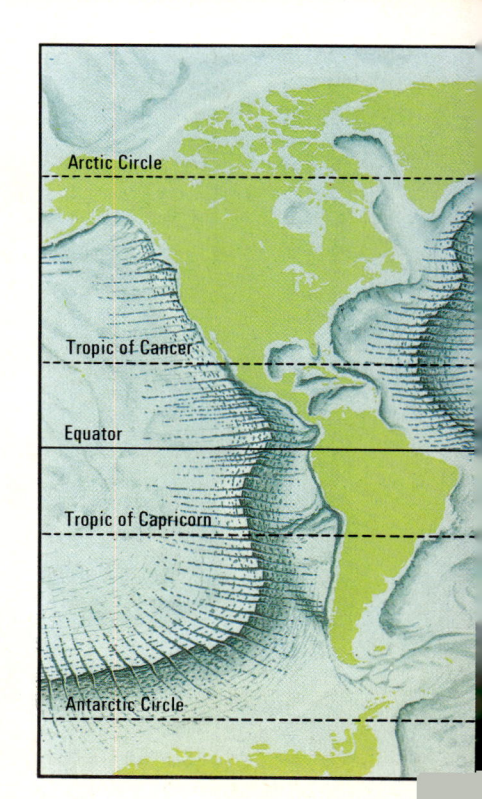

Arctic Circle
Tropic of Cancer
Equator
Tropic of Capricorn

Direction of plate movement
Possible plate boundary
Subduction zone
Spreading ridge
Collision zone
Volcano

Arctic Circle
Tropic of Cancer
Equator
Tropic of Capricorn
Antarctic Circle

12

How the Sea-Floor Is Made and Lost

By the 18th century, people had mapped most continental shores well enough to show a strange fact: if you could cut out the continents and push them together, their outlines would almost fit like the pieces of a jigsaw puzzle. (If you make their edges the rims of the offshore continental shelves, the fit is even better.)

In 1915 the German meteorologist Alfred Wegener suggested why. The continents, he argued, had once been joined, but gradually broke up and drifted apart. This theory of CONTINENTAL DRIFT suggested that the shapes and sizes of the oceans as well as of the continents are changing.

Half a century passed before scientists discovered and pieced together the clues that show how all this happened.

The continents and ocean floors are probably moved through the action of huge currents in the MANTLE – the thick semi-plastic layer beneath the Earth's crust. The crust is thinner under the ocean floor than under the continents. Where an upwelling current hits the ocean floor it divides and tears the seabed apart. Molten rock wells up to fill the gap. But the gap is always widening; thus as the new rock cools and hardens, it sticks to each side of the gap. This builds those great underwater mountain chains, the MID-OCEANIC RIDGES. The rift running down the middle of a ridge shows the gap where fresh molten rock is welling up. In this way new ocean floor is continually being made.

Proving the Theory

There are many clues to this process. For instance, the Mid-Atlantic Ridge has earthquakes, active volcanoes and submarine 'hot spots', but not the SEDIMENTS that would have settled on its rocks if they were old. Scientific proof that they are geologically young comes from analysing cores of seabed rock.

Evidence that the Atlantic seabed has been spreading lies in the north–south alignment of particles in its rocks. As molten rock hardens, its particles align with the Earth's magnetic field, but this has changed repeatedly since the Earth began. Changing patterns of alignment across the ocean floor prove that it has grown from the middle outwards. They prove, too, that no deep ocean floor is more than 200 million years old – about one-nineteenth the age of some continental rocks.

How the Sea-Floor Disappears

The reason why no part of the sea-floor is very old is simple: while it is being made in one place, it is vanishing elsewhere.

The Earth's crust consists of giant jigsaw pieces known as plates, which float like rafts upon the mantle. As spreading mantle currents drive two plates apart, they collide with other plates. In places this buckles the crust into mountain chains. Elsewhere the rim of one plate is driven down beneath another (maybe pulled down by a cooled, sinking CONVECTION CURRENT in the mantle). The result is a deep OCEANIC TRENCH. Here the oceanic crust is being gobbled up into the depths below.

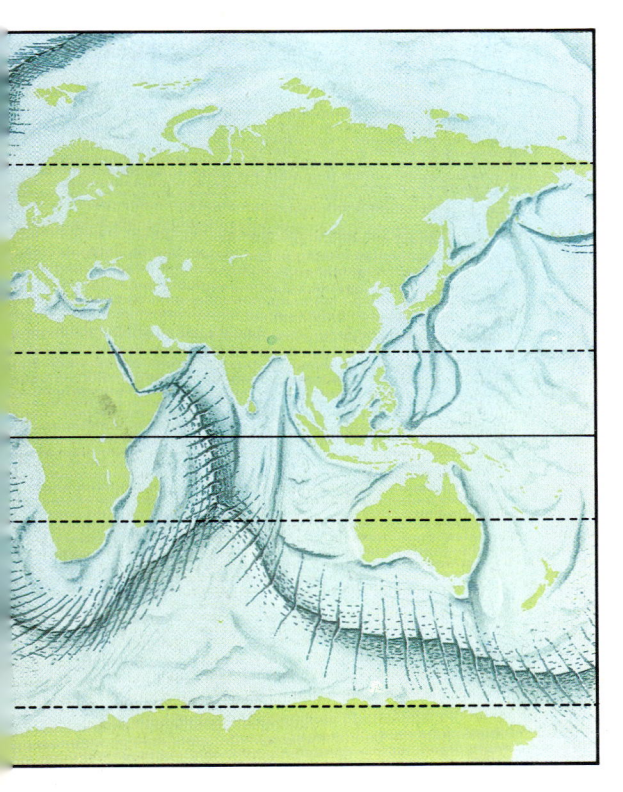

Left: This world map shows the vast submarine ridges that circle the oceans. A valley runs along the middle of each ridge, which is broken into sections by crosswise gashes known as transform faults.

Below: This section through the Earth's crust contrasts oceanic and continental crust. Oceanic crust is relatively thin, and has a simple structure. Continental crust is thicker and more complex. Both kinds of crust are less dense than the mantle rock on which they float.

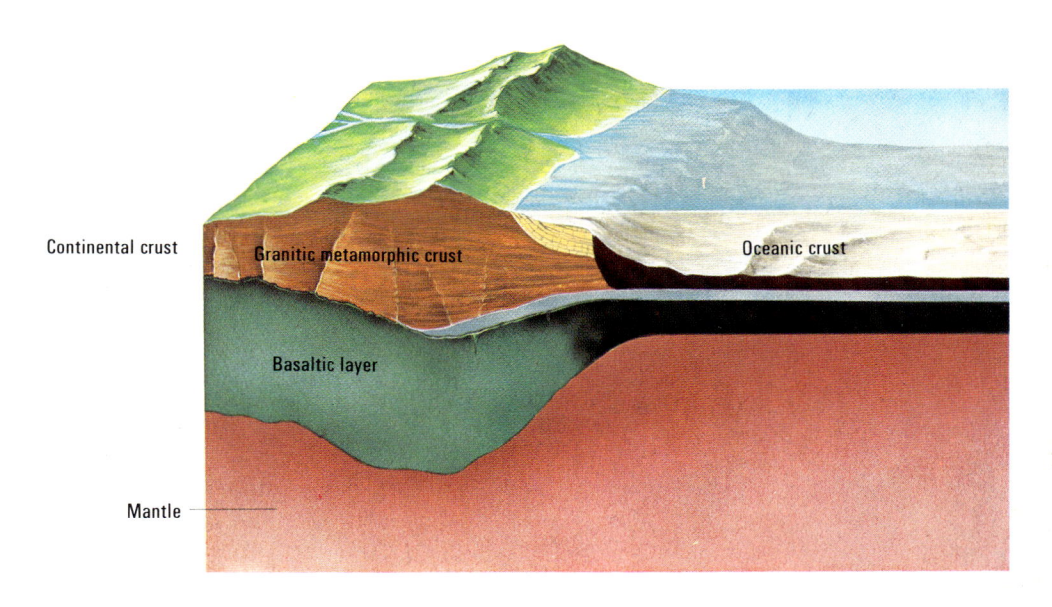

Submarine Landscape

Wind, rain, sun and frost sculpt the surface of the land. No such agents work upon the seabed. Yet this too has its mountains, plateaux, plains and valleys. Oceanographers often divide the ocean bed into three: the continental shelf, the continental slope, and the deep ocean floor.

The CONTINENTAL SHELF – the true edge of the continents – is a platform extending from low tide level to about 180 metres (600 feet) down. But its length varies enormously. It extends 1200 kilometres (750 miles) from Siberia into the Arctic Ocean. Off Chile, the shelf is almost absent and the coast plunges steeply into the depths.

Nowhere is the shelf entirely smooth. Submerged cliffs, beaches and river beds show where parts of the shelf once stood above dry land. Storm waves, rivers and tidal currents here and there heap TERRIGENOUS DEPOSITS (land-based deposits) of gravel, sand and mud into thick beds or underwater ridges that resemble dunes.

The Great Abyss

The gentle gradient of the continental shelf ends abruptly at the CONTINENTAL SLOPE. This inclines at an angle of between 4 and 20 degrees from the edge of the continental shelf to a depth of about 3800 metres (12,000 feet). The result is the world's longest and highest 'boundary wall'. In places deep gorges pierce its surface. Earthquakes opened up some gashes. Others mark the path of submarine silt-and-water avalanches. Some of these sudden falls have snapped underwater transatlantic cables.

Sediments collect at the bottom of the continental slope, forming the gentler slope of the continental rise. At the foot of this begins the deep ocean floor, the vast abyssal zone which, with its mid-oceanic ridges, makes up over half the Earth's total surface.

Underwater peaks and troughs remind us of the upwellings of molten rock that formed the ocean floor. Seamounts, or submerged volcanoes, thrust up from the lower depths. Hundreds have bevelled tops. These so-called GUYOTS once stood above the sea, but waves eroded and submerged them.

In contrast to the seamounts and the major mid-oceanic ridges are the deep narrow trenches with V-shaped cross-sections. Many lie off rows of volcanic islands (the West Indies, Philippines, Marianas). These in turn lie close to larger land areas.

In places, TURBIDITY CURRENTS have dumped vast loads of sand and mud, creating smooth abyssal plains. But the rain of debris on the open floor is generally thin. Sharks' teeth, volcanic dust and other strange ingredients occur in red clay. Dead planktonic plants and animals create pelagic OOZES. But in the Pacific Ocean such deposits build up at the rate of only 2·5 centimetres (1 inch) per 20,000 years. Meteorites add a small amount to the substances on the seabed. Each year up to 4 million tonnes of material from space fall on the Earth – mostly into the oceans.

The ocean bed. From a shallow continental shelf, a steep continental slope leads down to an abyssal plain, gashed by even deeper trenches. The vertical scale is exaggerated.

▓	Terrigenous deposits
▓	Calcareous ooze
▓	Red clay
▓	Radiolarian ooze
▓	Diatom ooze

Sediments covering the ocean floor. Terrigenous deposits include silt, mud and sand washed off the land. Red clay found in the deepest parts of the ocean, consists mainly of very fine particles. Calcareous ooze com... plankto... diato... to... a...

...f dead
...n and
...skele-
...and

Looking at the Oceans

The science of OCEANOGRAPHY, the study of the oceans and the sea-floors, is only a hundred years old. It began in the 1870s when HMS *Challenger* made a three-year world voyage to sound and sample the oceans and their contents.

The illustration below shows some of the many instruments now used to study the oceans.

Research ships check the salinity of the water at different depths by lowering valved Nansen bottles down a wire. Reversing thermometers attached to Nansen bottles take and record temperature at specific levels.

A bathythermograph, towed at depth behind a ship, continuously charts temperature and pressure. Free-fall recording devices find the speed and direction of underwater ocean currents (see page 22). Weighted floats emitting sonar signals can also be tracked as they drift in the currents, as can meters tethered to the seabed and then released by sonar signal from the surface.

In 1855, the American naval officer Matthew Fontaine Maury published the first map of an ocean bed. In those days, ships plumbed the depths haphazardly using a lead and line, and giving only sketchy information about the ocean floor.

Sea-floor exploration in shallow waters is now aided by powered submersibles (see page 55), and special craft have provided a firsthand glimpse of the abyssal floor. In 1960, Dr Jacques Piccard and Lt Donald Walsh in the BATHYSCAPHE *Trieste* plunged 10·9 kilometres (6·8 miles) to settle on the bottom of the Challenger Deep – the lowest place upon Earth's crust. Unlike ordinary submarines, *Trieste* was built to travel only up and down.

Most deep-sea seabed studies now use unmanned submersibles worked by remote control from a ship on the surface.

Above: HMS *Challenger*, from a painting in **1889.** In the **1870s,** *Challenger* sailed round the world studying the ocean bed, the water and sea life: the science of oceanography had begun.

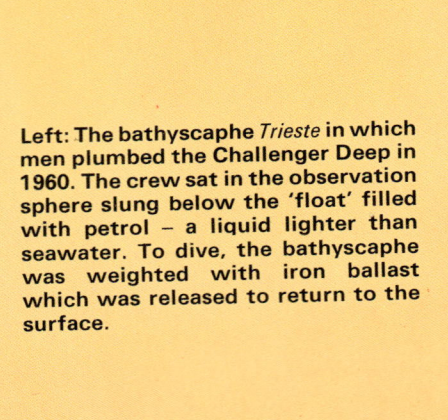

Free-fall sampler samples seabed sediment

Above: The *Glomar Challenger* **can drill into seabed 6·4 kilometres (4 miles) below the ocean surface.**

Left: The bathyscaphe *Trieste* **in which men plumbed the Challenger Deep in 1960. The crew sat in the observation sphere slung below the 'float' filled with petrol – a liquid lighter than seawater. To dive, the bathyscaphe was weighted with iron ballast which was released to return to the surface.**

Special Instruments

Underwater cameras let down by line photograph and film the floor of the abyss. Strong casing protects them from the great deep-sea pressure. Bottom-seeking pingers tell them when to work. Electronic flash equipment provides light. Special cameras take stereoscopic photographs or television film to show features such as seabed creatures, ripple marks and manganese nodules (see page 61).

The main snag with such devices is poor visibility due to particles suspended in the water. ECHO-SOUNDERS overcome this problem. These devices send out sound pulses and visually display the echoes reflected back from the seabed. An echo-sounder on a moving ship can show the depth of water extremely accurately down to 5500 metres (18,000 feet), and thus help hydrographers to plot seabed profiles.

Side-scan SONAR involves sound pulses sent down slantingly from a ship or sonar fish – an instrument-carrying 'torpedo' towed behind a vessel. The received echo pattern is built into a sonograph (an image traced by stylus on paper). This may show a clear photographic image of the seabed up to 24 kilometres (15 miles) away.

Sonar also helps us to discover what lies beneath the surface of the ocean floor. Acoustic pulses of high frequency and low energy penetrate sediment and probe rocks thousands of metres below the ocean bed.

Seismic profiling is another way to explore below the seabed. A survey ship sets off explosions under water, often in the form of shots from an airgun. HYDROPHONES towed by the ship analyse the echoes sent back by rocks at different depths.

Actually to examine seabed samples means bringing them to the surface. Grabs and dredges sample the upper seabed in ways much like those used by oceanographers a century ago. But modern drills and corers penetrate deep down. Corers are hollow tubes plunged in the floor to trap cores of layered sediments. *Glomar Challenger* has drilled 1300 metres (4265 feet) below the seabed.

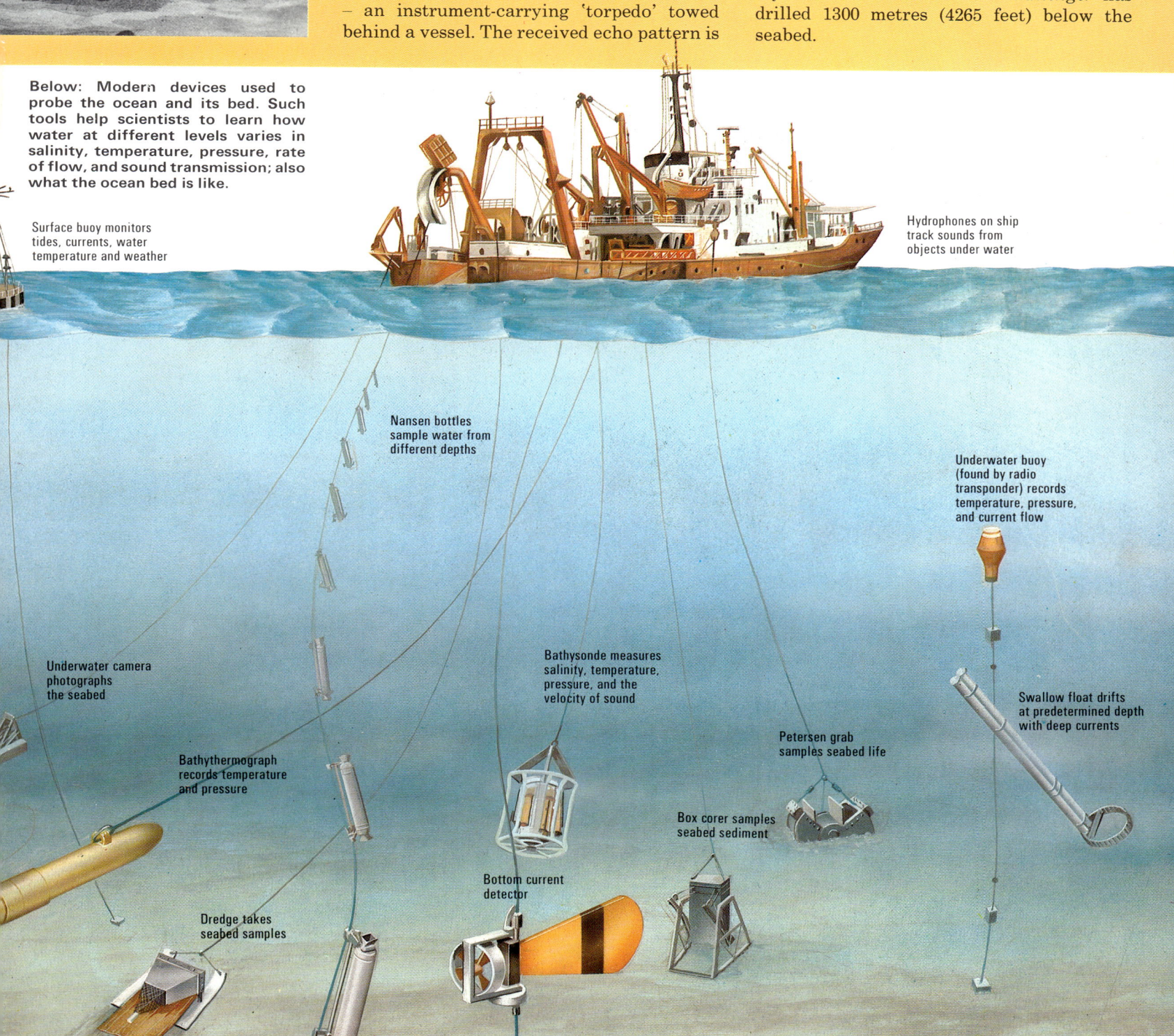

Below: Modern devices used to probe the ocean and its bed. Such tools help scientists to learn how water at different levels varies in salinity, temperature, pressure, rate of flow, and sound transmission; also what the ocean bed is like.

Surface buoy monitors tides, currents, water temperature and weather

Hydrophones on ship track sounds from objects under water

Nansen bottles sample water from different depths

Underwater buoy (found by radio transponder) records temperature, pressure, and current flow

Underwater camera photographs the seabed

Bathythermograph records temperature and pressure

Bathysonde measures salinity, temperature, pressure, and the velocity of sound

Petersen grab samples seabed life

Swallow float drifts at predetermined depth with deep currents

Box corer samples seabed sediment

Bottom current detector

Dredge takes seabed samples

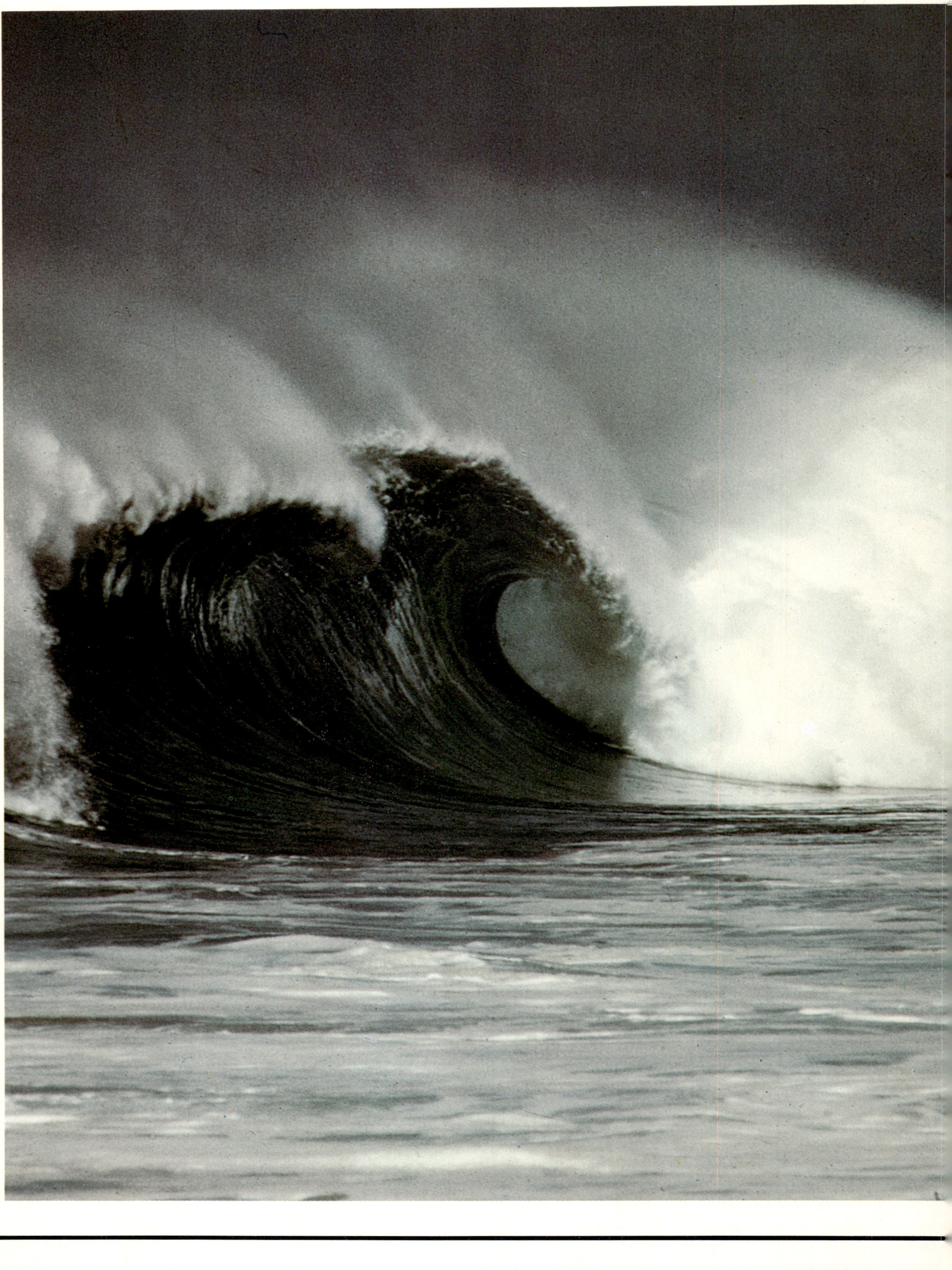

Chapter Two

THE OCEAN SURFACE

Unlike the surface of the land, the surface of the sea is continuously moving. Winds which blow steadily from one direction set up great currents through the sea. Sudden storm winds whip up the surface into waves as high as houses. The Earth's spin and the force exerted by the Sun and Moon combine to create the tides. Over thousands and millions of years, earthquakes and changes in climate have helped to make drastic changes in the level of the sea and land in relation to each other. Where the sea freezes, its surface forms slabs of ice that grind together.

   We are beginning to come to terms with the sea's uncertain surface. Ships use this water barrier as a highway. People are harnessing the energy locked up in waves and tides, planning to use icebergs as floating reservoirs, and in places even pushing back the edge of the ocean.

A breaking wave reveals the force and fury of surface waters when strong winds lash them into motion. Ice, currents, tides, and earth movements between them can also influence the form, flow or level of an ocean surface.

Ships

Ships have changed the oceans from barriers into highways. Early explorers of the sea – like the Vikings, the Polynesians and the Portuguese – travelled in small sailing boats and set up trade routes. But the invention of steam-powered vessels in the last century caused a great expansion in the volume of travel. Huge, steel-hulled, piston-engined steamships ousted the wooden sailing ships, and then themselves gave way to vessels where steam pressure spins turbine blades.

Electric motors, diesel motors, gas turbines and nuclear reactors are other systems of propulsion. In each case, engines thrust a vessel forward by spinning a propeller. Some propellers are housed in open-ended cylinders, or have blades that can be set at different angles. Devices like these make propellers more efficient and cut fuel costs. Other modern aids include underwater fins called stabilizers which reduce the ship's rolling motion in heavy seas, and the bulbous bow which reduces the bow wave that normally builds up and slows the ship down.

Ship Shapes
Vessels come in many shapes and sizes, depending on the task for which they are designed. The great ocean liners are now a thing of the past. Giants like the *France* and *Queen Elizabeth* were laid up or broken up when transoceanic emigrants and businessmen deserted them in favour of the aeroplane. But shipyards do build small luxuriously fitted cruise liners like the *Song of Norway*, with a yacht-like bow and an observation lounge.

Most sea passengers are now carried by the ferries and train ferries that cross such narrow water gaps as the English Channel and the Baltic Sea. These vessels have their rivals in the HOVERCRAFT, or air-cushion vehicles, and other small, fast craft called HYDROFOILS. Hydrofoils rise as they accelerate until they plane along on 'wings' which sprout from struts that jut down from the hull.

Cargo ships are by far the most important group of ships today. They carry the fuels, foods and manufactured goods that people need. Cargo ships of up to 12,000 gross tonnes (i.e. with about 34,000 cubic metres of permanently enclosed space) have several holds, or storage chambers. They handle a variety of cargo, and often have cranes on their deck to help them load and unload. Cargo liners make a regular run between specific ports, while so-called tramps sail on charter anywhere.

Container ships are cargo ships designed to bear containers of an internationally agreed size that may be loaded quickly in large numbers from lorries or railway wagons by means of dockside truck lifts and transporter cranes. Other major cargo ships include refrigerated fruit and meat carriers, bulk carriers and oil tankers. These last two ships have long low lines and superstructure near the stern. Bulk carriers take cement, coal, grain, ore and other bulky goods.

Oil tankers are specialized to take the liquid fuel without which world trade and industry would soon run down. By the 1970s, shipyards were delivering bulk carriers capable of carrying a quarter of a million tonnes, and supertankers with a capacity of more than half a million tonnes. By the end of the 1970s, however, many of these supertankers were lying idle. The sharp rise in oil prices meant that international trade in oil did not increase as much as people thought it would.

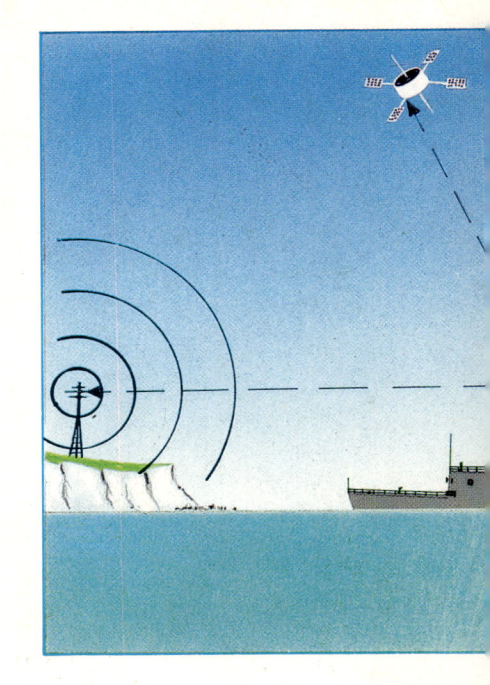

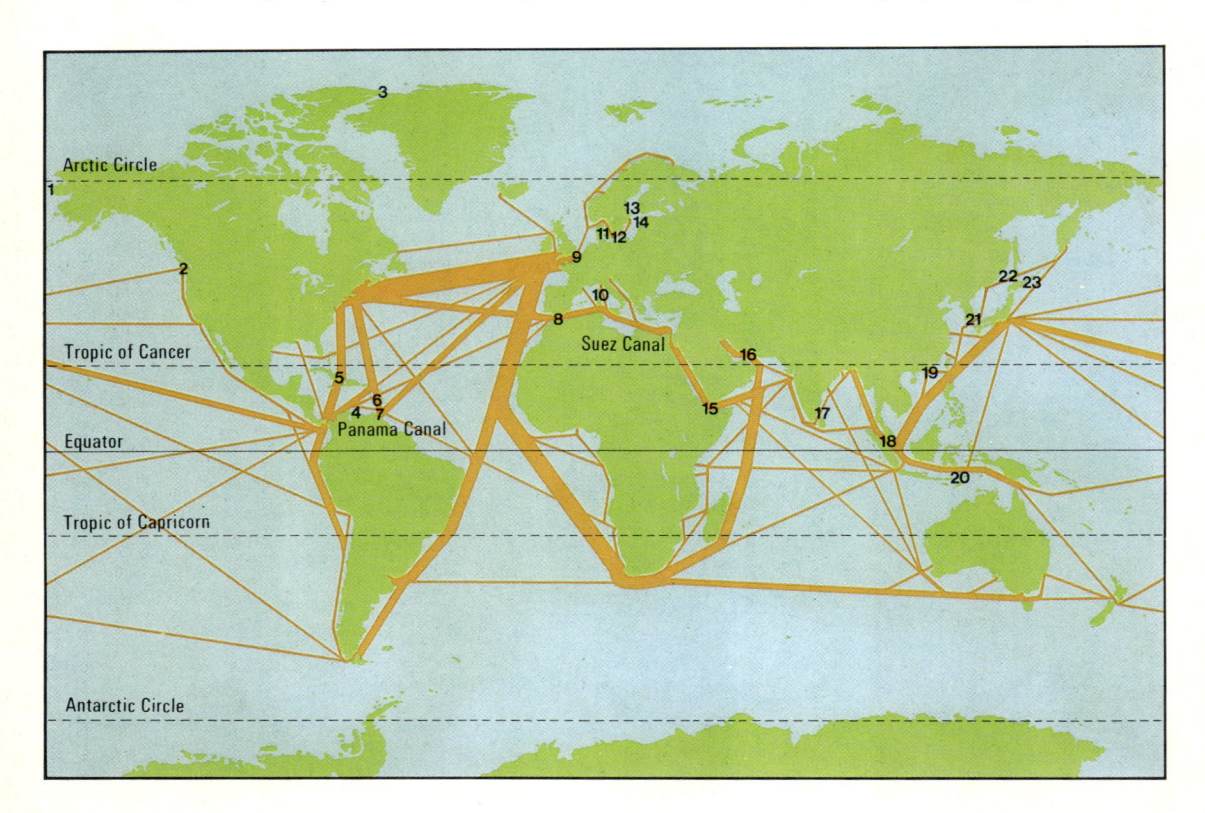

1 Bering Strait
2 Strait of Juan de Fuca
3 Robeson Channel
4 Aruba-Paraguana Passage
5 Windward Passage
6 Martinique Passage
7 St Lucia Channel
8 Strait of Gibraltar
9 Strait of Dover
10 Strait of Bonifacio
11 Öresund
12 Bornholmsgattet
13 Entrance to Gulf of Bothnia
14 Entrance to Gulf of Finland
15 Bab al Mandab
16 Strait of Hormuz
17 Palk Strait
18 Strait of Malacca
19 Lema Channel
20 Strait of Ombai, Selat
21 Chosen Strait
22 Soya Strait
23 Notsuke-suido

Left: Lines show world shipping lanes. Thick lines show the busiest ones. The numbers show international straits. Whether these straits are 'free' or controlled by the nations on each side is crucial to shipping.

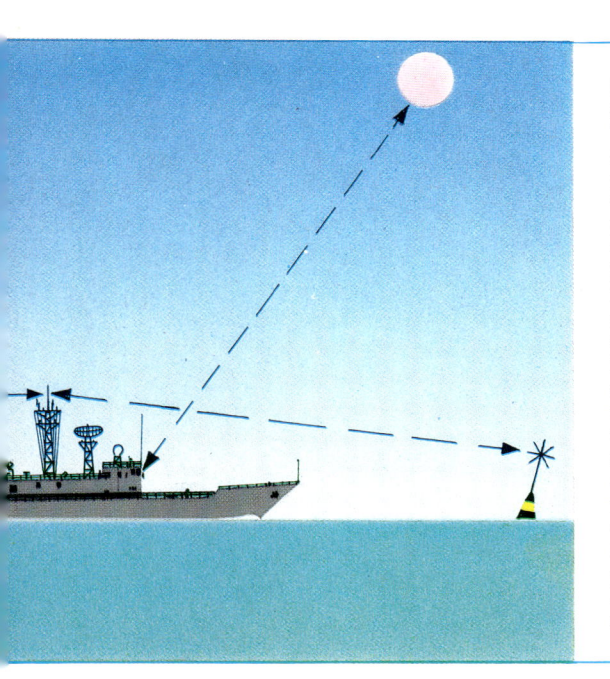

Left: Various navigation aids help ships to determine their position. They can check position in relation to certain artificial satellites; from radio signals beamed out by shore transmitters; or by radar echoes bounced off navigation buoys. Meanwhile gyrocompasses enable them to keep their headings in relation to true north. In the old days, ships used unreliable magnetic compasses; found longitude (east–west position) only by using clocks called chronometers; and used sextants to find the altitude of the Sun and certain stars and hence their own latitude (north–south position). Modern electronic aids give greater accuracy than older means of navigation, which still, however, have their uses, especially if other systems fail. Accurate navigation is vital – not just to keep a ship on course but to prevent it striking rocks or other ships.

Warships

Major powers build warships for two main reasons. They want to protect their merchant ships, shipping lanes, fishing grounds and offshore oilfields vital to their economies. They also use the oceans as a mobile base from which to threaten hostile countries with long-range missiles armed with nuclear warheads.

Radar-guided missiles have made big warships vulnerable and largely obsolete. A modern navy's strike force lies largely in its guided missile cruisers and destroyers, and in nuclear-powered submarines.

Below: Five contrasting vessels. Hovercraft speed passengers over narrow seas. Trimarans are multi-hulled sailing craft sometimes seen in ocean races. Tankers and container vessels shift vast loads of vital goods around the world. A destroyer typifies the modern warship. The ships are drawn to scale.

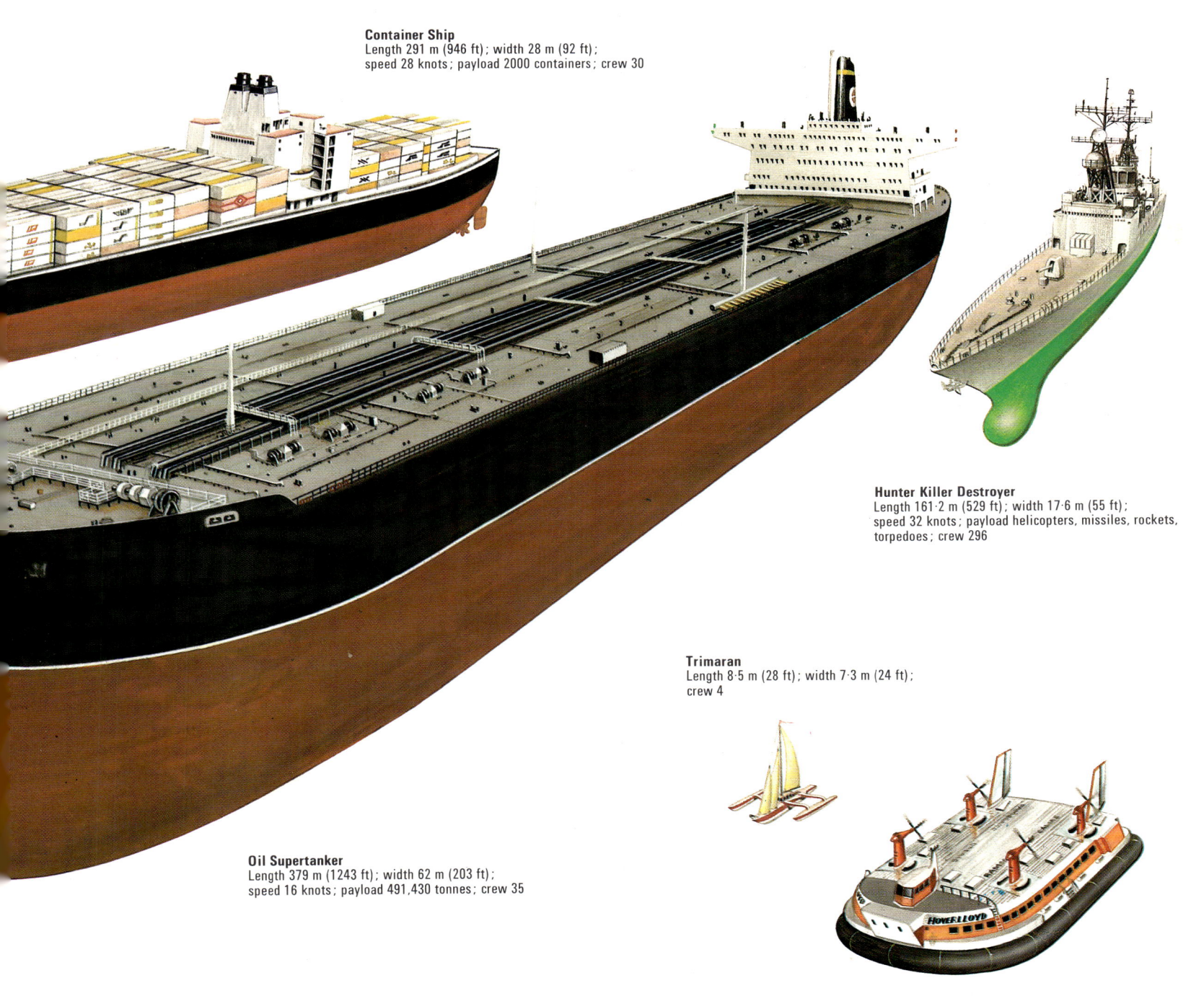

Container Ship
Length 291 m (946 ft); width 28 m (92 ft); speed 28 knots; payload 2000 containers; crew 30

Hunter Killer Destroyer
Length 161·2 m (529 ft); width 17·6 m (55 ft); speed 32 knots; payload helicopters, missiles, rockets, torpedoes; crew 296

Trimaran
Length 8·5 m (28 ft); width 7·3 m (24 ft); crew 4

Oil Supertanker
Length 379 m (1243 ft); width 62 m (203 ft); speed 16 knots; payload 491,430 tonnes; crew 35

Hovercraft SRN4
Length 39·7 m (130 ft); width 23·4 m (77 ft); speed 77 knots; payload 34 cars, 174 people; crew 11

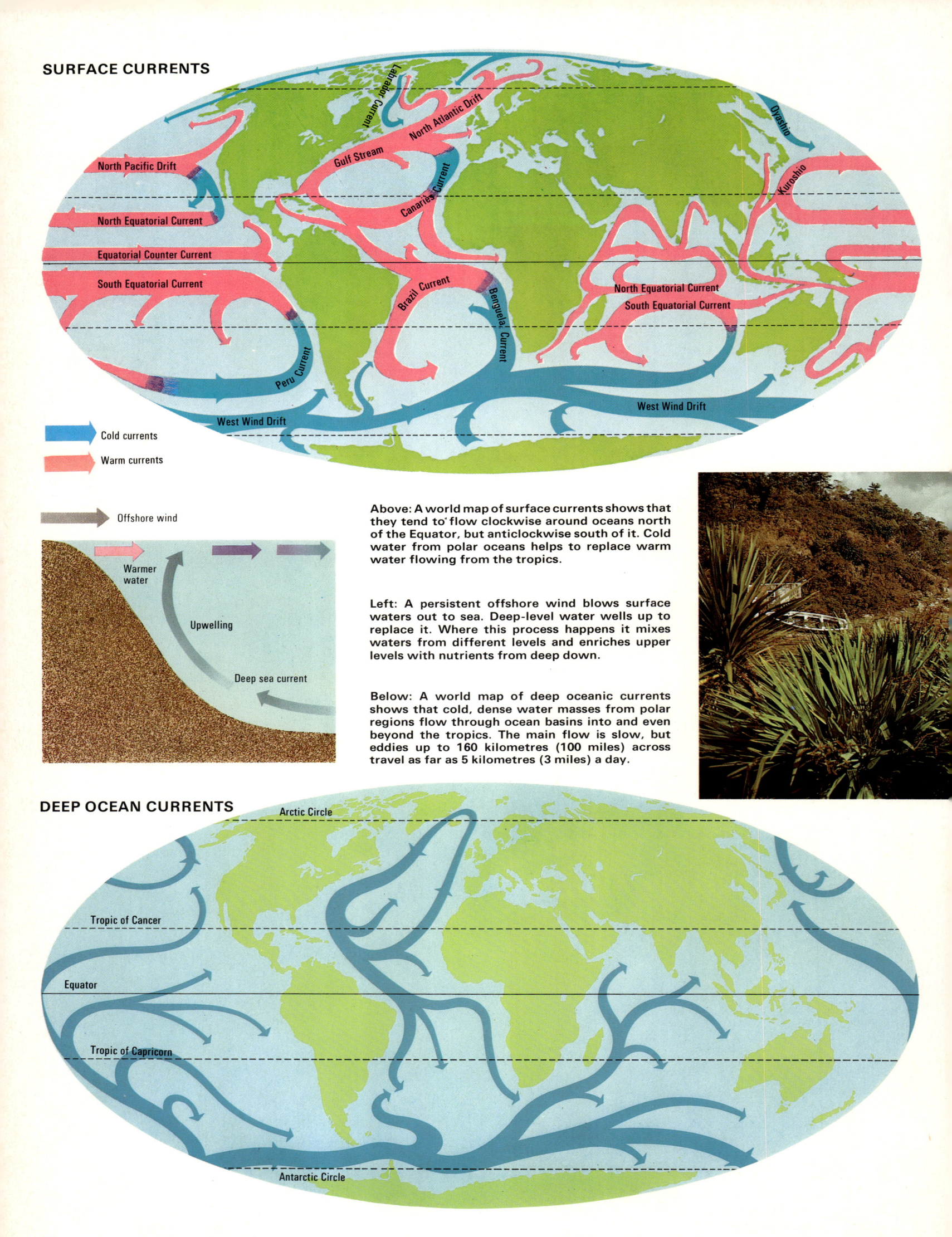

SURFACE CURRENTS

Labrador Current · North Atlantic Drift · Oyashio · Gulf Stream · North Pacific Drift · Kuroshio · Canaries Current · North Equatorial Current · Equatorial Counter Current · South Equatorial Current · North Equatorial Current · South Equatorial Current · Brazil Current · Benguela Current · Peru Current · West Wind Drift · West Wind Drift

→ Cold currents
→ Warm currents

→ Offshore wind

Warmer water · Upwelling · Deep sea current

Above: A world map of surface currents shows that they tend to flow clockwise around oceans north of the Equator, but anticlockwise south of it. Cold water from polar oceans helps to replace warm water flowing from the tropics.

Left: A persistent offshore wind blows surface waters out to sea. Deep-level water wells up to replace it. Where this process happens it mixes waters from different levels and enriches upper levels with nutrients from deep down.

Below: A world map of deep oceanic currents shows that cold, dense water masses from polar regions flow through ocean basins into and even beyond the tropics. The main flow is slow, but eddies up to 160 kilometres (100 miles) across travel as far as 5 kilometres (3 miles) a day.

DEEP OCEAN CURRENTS

Arctic Circle · Tropic of Cancer · Equator · Tropic of Capricorn · Antarctic Circle

Currents

CURRENTS are 'rivers in the sea' that flow through the surface waters of all major oceans. One of the forces that drive them is the wind. Close to the equator, the steady TRADE WINDS drive surface waters westward. Towards the polar regions, strong westerly winds drive the waters eastward.

When a current reaches a continent, the land mass forces it to alter course. Then, too, the Earth's spin deflects water flowing from the equator towards the poles. This tendency is called the CORIOLIS EFFECT, after its discoverer (in 1835), the French mathematician Gaspard de Coriolis. North of the equator the Coriolis effect bends currents to the right; south of the equator, to the left.

Direction of currents

Between them, trade winds, westerlies, land masses and the Coriolis effect mean that surface currents loop clockwise around oceans in the northern hemisphere. Oceanographers call these looping current systems GYRES.

In the North Atlantic gyre, for instance, the North-East Trade Winds trigger the westward flowing North Equatorial Current. Funnelled through the Florida Strait, the moving water emerges as the Gulf Stream. This warm current, up to 60 kilometres (40 miles) wide and 600 metres (2000 feet) deep, snakes northward at 37 kilometres (20 miles) a day off eastern North America. South-westerly winds then push it north-eastward; it broadens, slows down and divides. As the North Atlantic Drift, part of the weakened current wafts warm water to north-west Europe. Reinforced by cold upwelling water, a southern arm bends south to become the Canaries Current. This flows off north-west Africa, and rejoins the North Equatorial Current. Still waters inside the North Atlantic gyre make up the weedy tract of ocean called the Sargasso Sea.

Of course, the full North Atlantic current system is more complex. For example, the Labrador current washes North America's north-eastern coasts with chilly polar water.

Effects of currents

Because they affect the air above them, warm and cool surface currents also affect the climates of the areas nearby. Thus the warm North Atlantic Drift and North Pacific Drift keep Norwegian and Alaskan ports respectively ice-free in winter. Meanwhile coasts much farther south (but on the *eastern* coasts of Eurasia and North America) are icebound – largely thanks to the cold Oyashio and Labrador currents.

Currents also shift plants and animals around. Many Welsh babies once teethed on tropical beans washed in by ocean currents. Nipa and other coconut palms sprout on Pacific shores where winds and currents dump their germinating fruits. Baby eels, hatched in the Sargasso Sea, hitch a ride to Europe with the North Atlantic Drift.

Deep Sea Currents

Most surface ocean currents peter out by 200 metres (656 feet) down. But oceanographers now know that sluggish counter currents flow far beneath the surface. Cold, dense polar waters sink and spread along the ocean floor towards the tropics. Where winds blow surface waters from the coast, deep, cold waters well up to take their place. UPWELLING of this kind explains the colder Peru (or Humboldt) Current and Benguela Current flowing north, respectively, off south-west South America and south-west Africa.

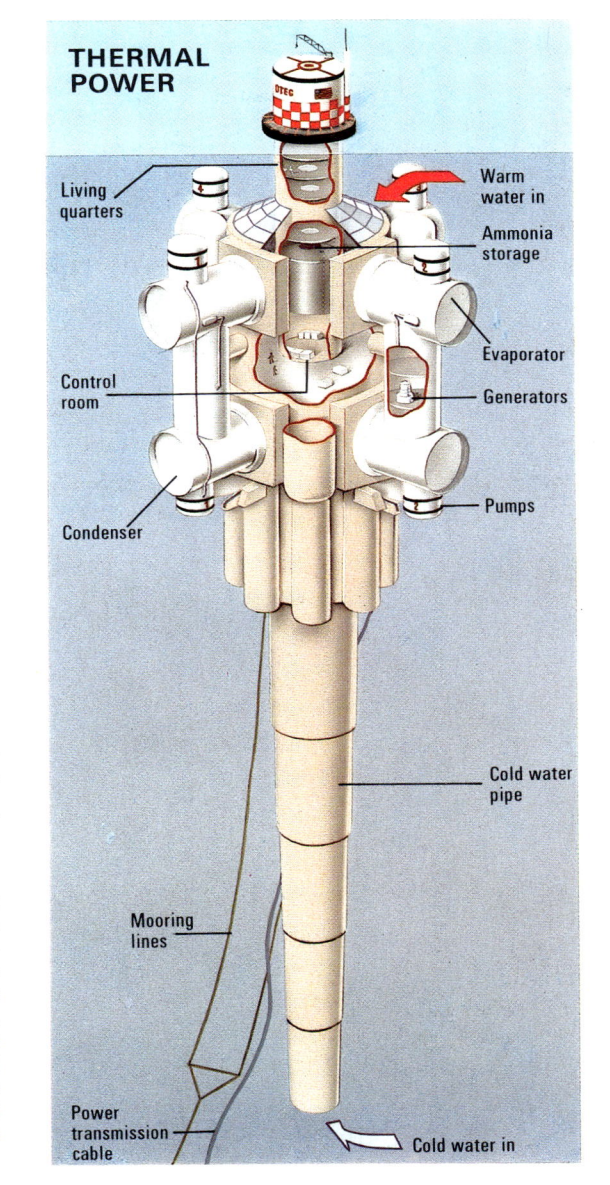

Below: These palm trees recall the tropics but grow almost as far north as Moscow. Here, in the Western Highlands of Scotland, the North Atlantic Drift warms the air and winter frosts are rare.

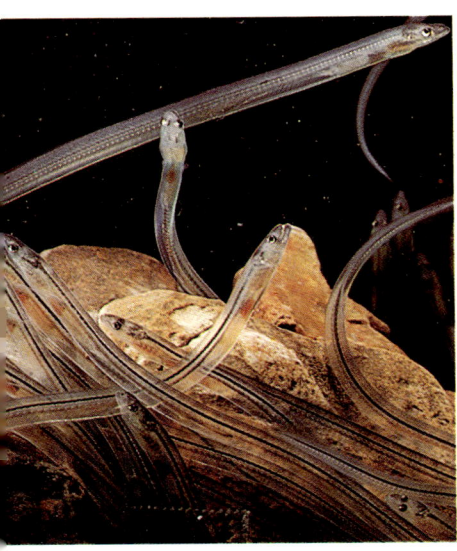

Above: Young eels at this stage of growth are called elvers. They are hatched in the Sargasso Sea, and travel across the Atlantic with the North Atlantic Drift. Two years later they reach Europe.

Right: Giant floats like this may use the fact that deep water in most oceans is colder than that at the surface. In this OTEC (Ocean Thermal Energy Conversion) device, pumps draw in warm surface water to vaporize ammonia. The vapour spins turbine blades and generates electric current. Next, near-freezing water from deep down is drawn up through a central pipe to cool and condense the vapour for re-use. Such a structure would have four generating plants and be 500 metres (1640 feet) high. It would work best in the tropics.

THERMAL POWER — Living quarters, Warm water in, Ammonia storage, Evaporator, Control room, Generators, Condenser, Pumps, Cold water pipe, Mooring lines, Power transmission cable, Cold water in

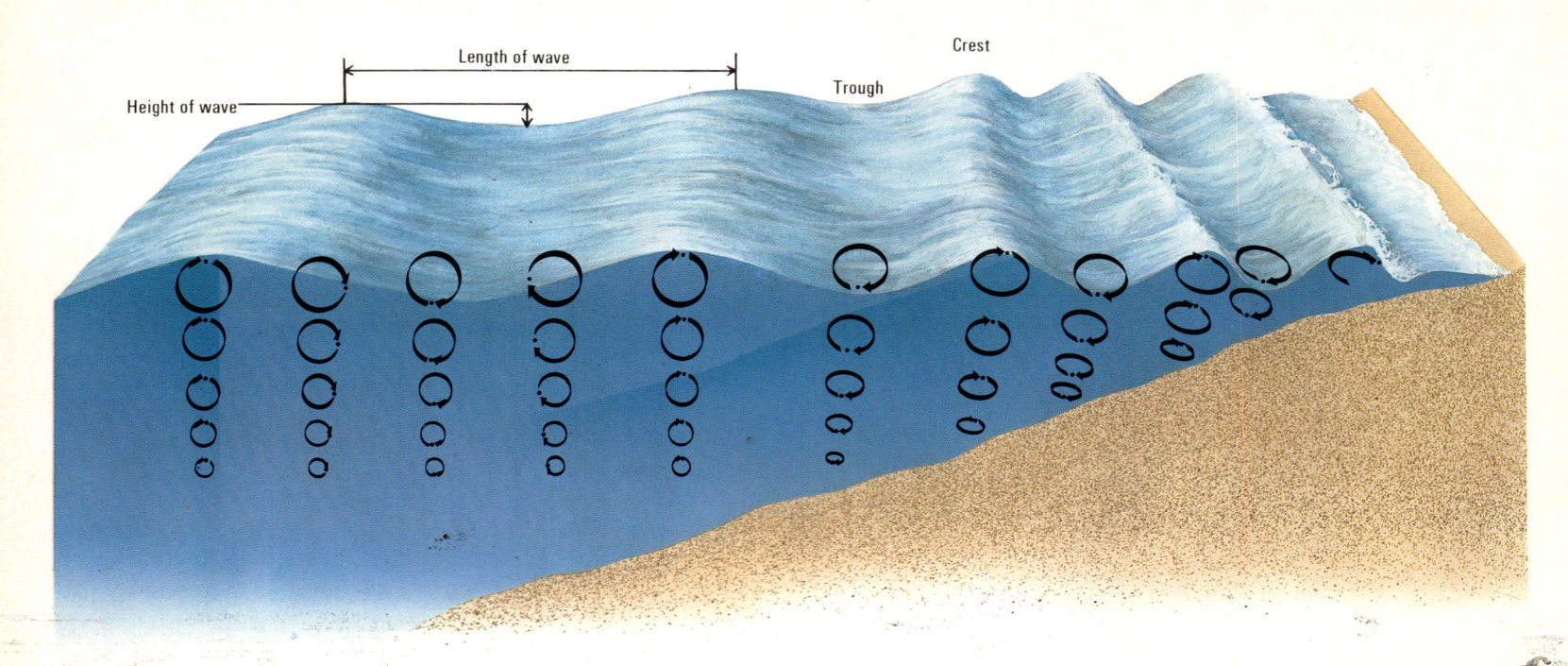

Waves

The sea is seldom truly flat and calm. Its surface is usually agitated and thrown up into rows of moving ridges and valleys. Small agitations of this kind are ripples; the larger ones are WAVES. Both are set in motion by the wind. The stronger and more persistent the wind, the larger the waves. Their size increases, too, with the FETCH – the length of open sea over which the wind is blowing. The Pacific, with the greatest fetch of any ocean, has the largest waves of all.

Waves build up as wind speed rises. Winds of 2·5 kilometres (1·6 miles) per hour start ripple patches known as 'catspaws' ruffling the surface. Winds of 13 kilometres (8 miles) per hour build catspaws into steep waves with breaking crests. As wind speed goes on rising, waves grow in height and length. Waves called SWELLS and measuring 1 kilometre from crest to crest may move at 55 kilometres (34 miles) per hour. Giant waves as high as 34 metres (112 feet) can occur where storms drive waves from the deep, open ocean into the shallow waters off a continent.

If unobstructed by land, waves whipped up by storms will travel far beyond their source. Waves generated in the Indian Ocean have travelled 19,000 kilometres (12,000 miles) and reached Alaska.

Tsunamis

Storm waves can cause extensive damage, but the most terrifying waves of all are TSUNAMIS (commonly, though inaccurately, called tidal waves). These are unleashed by submarine earthquakes or explosive volcanic eruptions. Each wave can cross an ocean at 500–800 kilometres (300–500 miles) per hour. A series of tsunamis 160 kilometres (100 miles) apart may be set off at 15-minute intervals. In the open ocean their amplitude is so low that they pass ships undetected. But tsunamis funnelled into shallow shores may tower 67 metres (220 feet) and devastate low coasts. In 1883, tsunamis from erupting Krakatoa killed more than 36,000 Indonesians. Some 300 tsunamis have been recorded during the last 2500 years – most originating in or near the Pacific Ocean.

What Happens in Waves

Out in the open ocean, the water does not travel with the waves that agitate it. Instead, each passing wave crest lifts water particles that then move forward, down, and back again, completing a circle. This is why seagulls and buoys floating on the open sea just bob up and down as waves pass by. Each circling motion of the upper water particles triggers a stack of smaller circlings in the water particles below it.

However, something different happens when waves reach shallow water – water less than half a wave-length deep. Water particles circling at the lowest levels drag on the seabed. This slows the waves. Their crests crowd close together, and they pile up and break on the shore.

Above: Why waves break inshore. Passing waves set stacks of water particles circling. In water shallower than half a wavelength (distance crest to crest) circles flatten out into ellipses, waves slow down and crests pile up, then break as surf.

Right: This wave may have started in mid ocean as a gentle undulating swell. As it neared the coast its length diminished, its height increased, and it became an unstable breaker bursting on the shore.

Below: A 19th-century artist's impression of the tsunami that struck the Portuguese capital Lisbon in November 1755. One of the most devastating seismic waves recorded, it struck the lower levels of a city already being toppled by the shock of the earthquake. Between them, tsunami and shock killed over 20,000 people. Tsunamis happen when the seabed suddenly moves up or down causing an abrupt change in the water level above it. This generates giant, high-speed waves that may do more damage than the largest waves whipped up by storms.

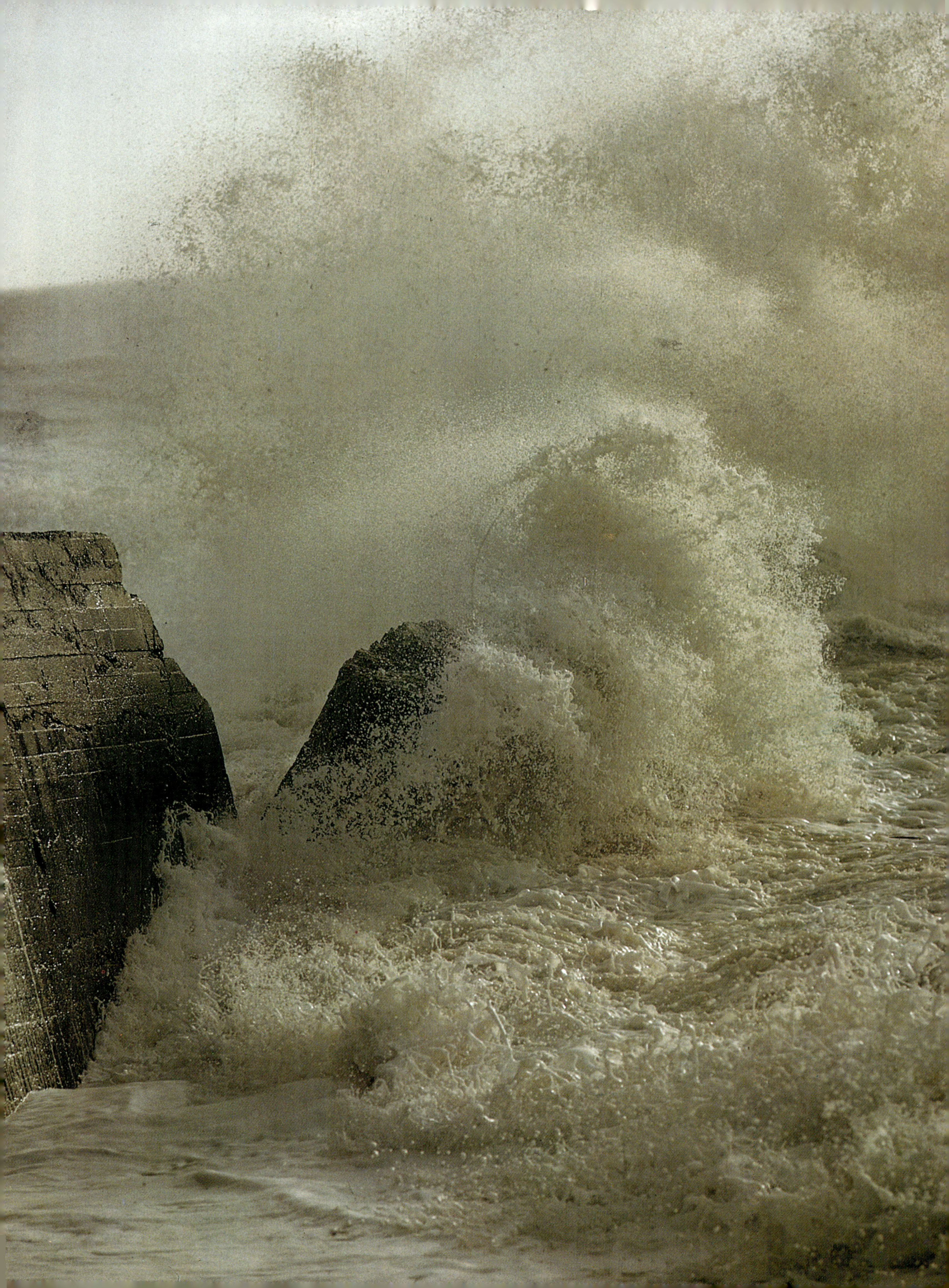

Wave Energy

Each day waves batter the world's shores with a total force equivalent to a 50-megaton hydrogen bomb. So far the colossal energy in wind-generated waves is wasted. But scientists have plans to harness WAVE ENERGY to work machines that generate electric current.

By the late 1970s several schemes were being tested. One involved a 'nodding duck' or rocking boom device. Passing waves set the duck nodding relative to the central spine, and devices in it convert the movement to electrical energy. Giant ducks arranged in rows the size of supertankers and strung out across 1000 kilometres (620 miles) of the North Atlantic could theoretically supply Britain with much of the energy it needs.

Contouring rafts is another scheme. Waves jostle the rafts and their motion operates hydraulic pumps hinged between them, converting the energy into high pressure in a fluid.

A third invention – the oscillating water column – works on the principle that the wave motion inside an inverted box can force air in and out of a hole in the top.

A fourth scheme – the Russell rectifier – looks like a huge floating box comprising reservoirs at different levels. Waves force water into upper reservoirs and out of lower ones to generate electric current.

Huge problems must be solved before any of these schemes proves practicable. Vast floating generators would pose hazards to shipping, and might break loose in storms unless tied down as securely as an oil rig. Nonetheless, a looming shortage of oil makes their development urgent.

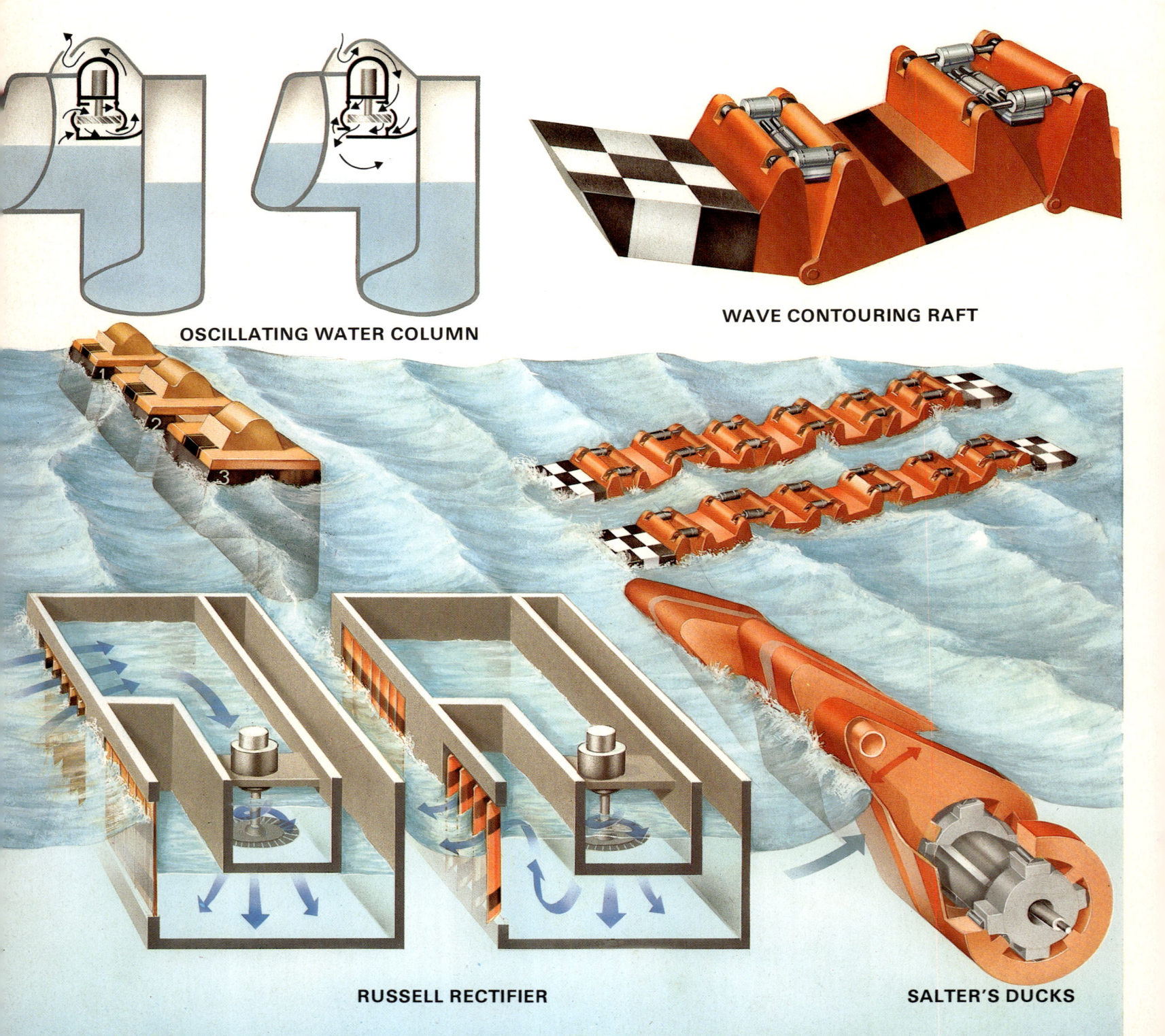

OSCILLATING WATER COLUMN

WAVE CONTOURING RAFT

RUSSELL RECTIFIER

SALTER'S DUCKS

Tides

Half the fun of building sandcastles for children is watching the rising tide wash them away. On most shores the sea rises and falls twice in every 24 hours 50 minutes. This timing coincides with one orbit of the Moon around the Earth, because TIDES are chiefly caused by the Moon's gravitational force pulling on the surface of the Earth.

As the Moon tours around the Earth it tends to pull towards it those oceans on the side of the Earth facing the Moon. At the same time the Earth's spin tends to push waters outward from the side of the Earth farthest from the Moon. In theory this produces two bulges or 'standing waves' that sweep around the Earth in alignment with the Moon. Each such wave represents a high tide; each 'trough' a low tide.

The Complicated Tides

In fact tides work in a more complicated way than this. First, the Earth's spin sets up friction that slows down and tends to deflect the tidal waves. Secondly, the continents and underwater ocean ridges prevent tidal waves surging freely all round the world. Thirdly, the Sun adds its own pull to the Moon's.

For these reasons, tides don't always happen when, where and how you might expect them. In fact the oceans are subdivided into 'tidal units'. In each unit, tides revolve about a so-called nodal point somewhat as a wobbling top spins about its centre.

At the nodal points themselves the tidal range is nil. Elsewhere in the open ocean it may be roughly half a metre. But as tidal waves sweep into shallow coastal waters they speed up, and pile up, reaching several metres above the level of EBB TIDE.

Some bays and seas adjoining oceans have especially high tides. This is because they 'bounce back' one high tide in such a way that the reflected tidal wave meets and reinforces the next one sweeping in. Such so-called 'resonant' tides occur in the North Sea and the English Channel. The world's highest tides happen where the Atlantic funnels into the long, narrow Bay of Fundy, on Canada's eastern coast. Here, water can rise to roughly twice the height of a house.

Seas like the Baltic and Mediterranean are at the opposite extreme. Almost shut off from the open oceans, they have a very low tidal range indeed.

Tides vary locally in frequency as well as range. Most shores receive two high and two low tides every lunar day. Atlantic coasts generally have 'semi-diurnal tides' (tides with level highs and level lows). But much Pacific and Indian Ocean shoreline has 'mixed tides' – tides with highs or lows that are unequal. Alaska, the Gulf of Mexico, the Philippines and part of China have only one high and one low tide per day – a system of 'diurnal tides'.

Spring Tides and Neap Tides

Tides vary with time as well as place. On any stretch of shore, SPRING TIDES (the highest high tides and lowest low tides) occur when the Sun and the Moon line up and pull together on the oceans. NEAP TIDES (the lowest high tides and highest low tides) occur when the Sun's pull is at right angles to the Moon's. Spring tides happen about twice a month, near the times of new moon and full moon. Neap tides coincide with the Moon's first and last quarters.

The world's tides contain a wealth of almost untapped energy. But in 1966 a barrage across France's Rance estuary began harnessing tides to generate electric current. Similarly the Bay of Fundy, the River Severn and the White Sea may one day provide energy for Canada, Britain and the Soviet Union respectively.

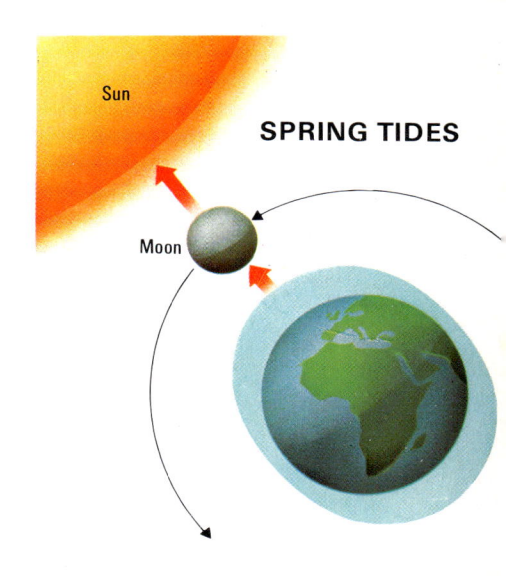

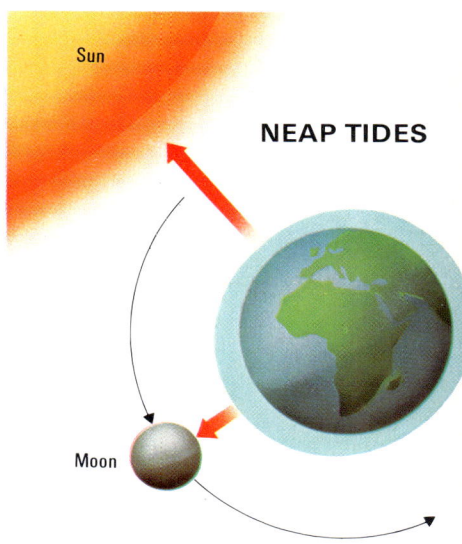

SPRING TIDES

NEAP TIDES

Above: Changing positions of the Sun and Moon cause spring tides at new (and full) moon; neap tides at half moon.

TIDAL POWER IN THE RANCE ESTUARY

A barrage across the Rance estuary in Brittany (below and below left) exploits a tidal range of 13.5 metres to generate 544 million kilowatt hours of electricity a year. To flow upstream the rising tide must pass and spin the blades of 24 turbines installed in the barrage (below). When the tide falls, the same thing happens in reverse. Extra power is gained by pumping more water in at high tide and out at low tide, so giving a subsequently greater difference in level.

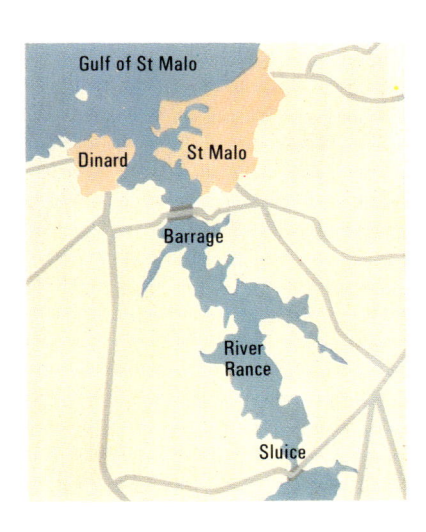

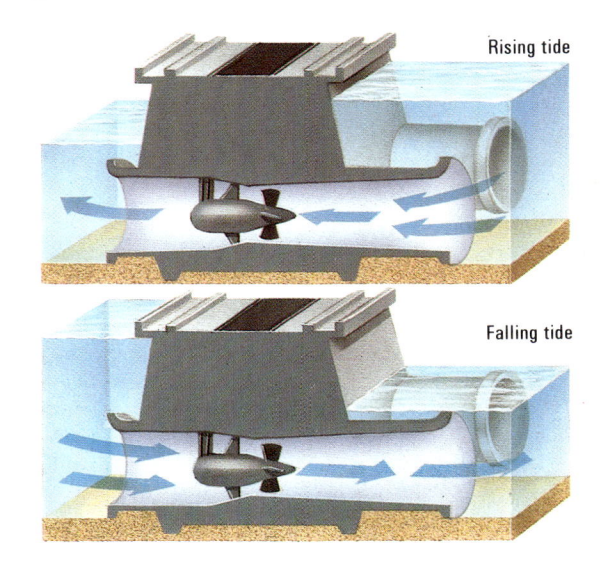

Rising tide

Falling tide

27

Changing Sea Levels

Left: A glacier and snowy peaks in Alaska recall the ice sheets that cloaked much of the northern hemisphere in the coldest phases of the Ice Age. Much water evaporating from oceans fell as snow on ice. Robbed of recycled water, the oceans shrank and their levels dropped.

The rise and fall of the tide is easy to see. Less obvious are far slower but far greater changes in sea level. Some occur because land sinks or rises. Changes involving a local land mass are called tectonic; those affecting a continent or region are called epeirogenic. Alterations in the level of the sea itself are called EUSTATIC CHANGES.

There have been many eustatic changes in the last million years or so as the extent of the ice caps has grown and shrunk during phases of the Ice Age. At one time so much water was locked up in the ice sheets that the level of the oceans was possibly some 90 metres (300 feet) below that at present. Huge tracts of offshore seabed became dry land. About 16,000 years ago it would have been possible to walk from Ireland to France. The city sites now on the USA's eastern seaboard then lay far inland. A mighty peninsula now known as Sundaland stuck out from southeast Asia, embracing the islands of Borneo, Java and Sumatra. The so-called Bering Land Bridge between Alaska and Siberia once linked North America and Asia.

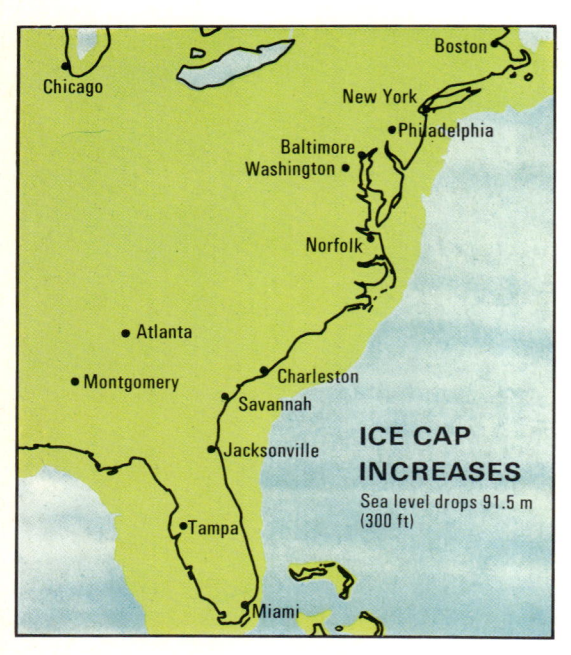

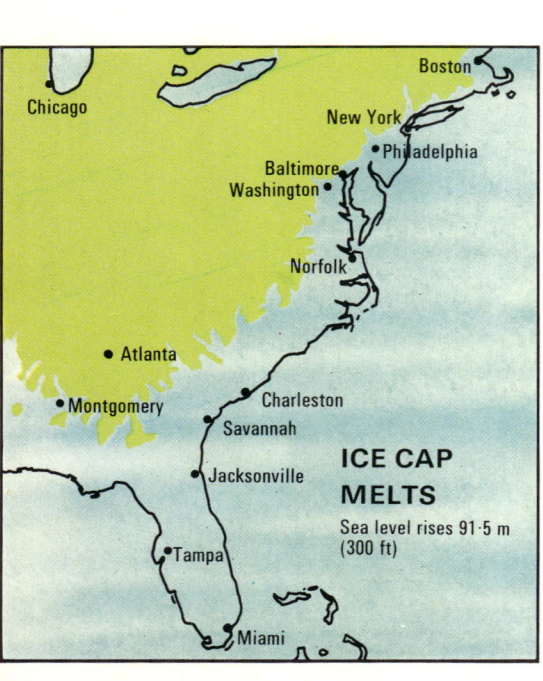

Drowning Lands

The great ice sheets covering much of northern North America and Asia began to melt about 10,000 years ago and flow into the oceans, whose level rose dramatically. In northern Europe a great low-lying plain was drowned, producing the North Sea. In time the invading sea isolated Ireland from Great Britain, and Great Britain from the rest of Europe. All but the highest parts of Sundaland became submerged.

Left: How climate changes could affect the eastern US coastline. In intense glaciation, sea level might fall 91·5 metres (300 feet), stranding ports inland. Melting of all polar ice could raise sea level 91·5 metres (300 feet), drowning great tracts of coastal plain including Florida.

ICE CAP INCREASES Sea level drops 91.5 m (300 ft)

ICE CAP MELTS Sea level rises 91.5 m (300 ft)

THAMES BARRAGE London is built on clay and is slowly sinking. A tidal barrage like that shown in the model is being built across the Thames to save large areas of London from possible flooding. The diagram shows how the flood gates can be rotated and closed against dangerously high tides.

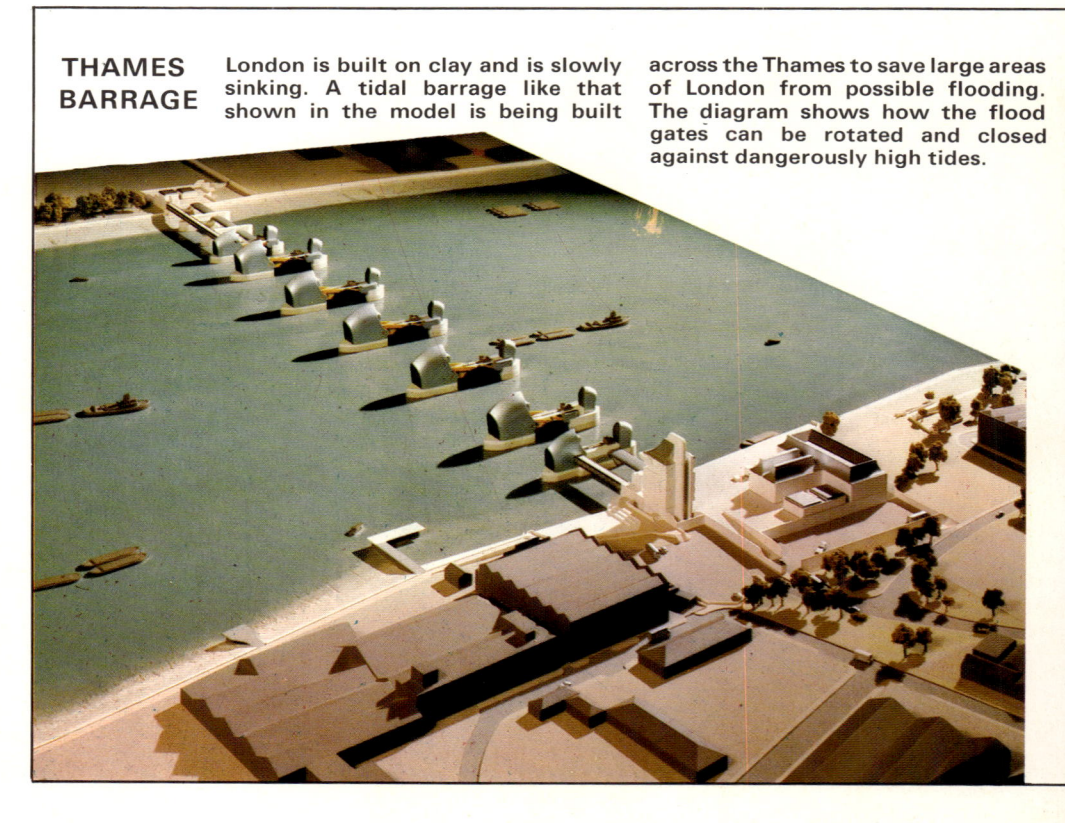

Coastlines as we know them took shape. Many show tell-tale signs of having been submerged in the past. Where the sea invaded a hilly or mountainous area, it created a ragged coast. Old valleys became inlets, and old uplands became peninsulas and islands. Where such hills and valleys jut straight out into the sea, broad, funnel-shaped estuaries called RIAS eat deep into the land. Devon, Cornwall, south-west Ireland, western Brittany and north-west Spain all have such ria coastlines.

Where the sea invaded a valley deeply gouged out by a glacier, a long, steep-sided inlet called a FJORD resulted. Fjord coastlines occur in Alaska, British Columbia, Labrador, Norway, Scotland, Chile and New Zealand.

As the sea spilt into mountain valleys that ran parallel to the coast, seaward ranges were transformed into long, narrow offshore islands alongside the shore. Coasts like this are features of Dalmatia in Yugoslavia and western North and South America.

Drowned lowland coasts are different. Here, river valleys expanded into broad estuaries, where low tide bares extensive mud flats and, in places, the remains of long-drowned forests. Marshes, spits and lagoons occur on such shores, too.

Raised Coastlines
In some places, the shoreline becomes higher than the sea, either because the sea level drops or because the land itself rises. If an ice sheet melts, the land below, which had been thrust down by the weight of the ice, starts an isostatic 'bouncing up'. Northern Scotland and Norway, for example, are still rising following the last glaciation. Here you may find raised wave-cut platforms, sea cliffs, and raised beaches left high and dry inland.

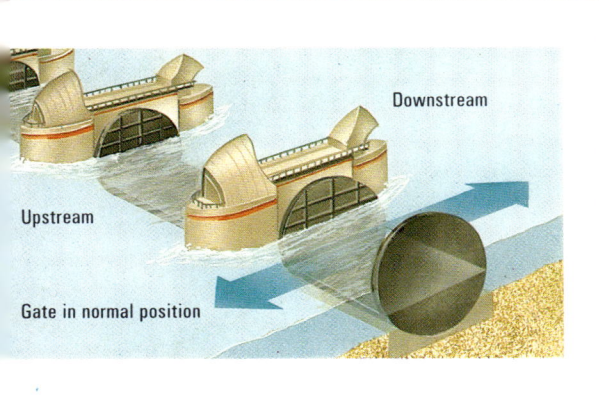

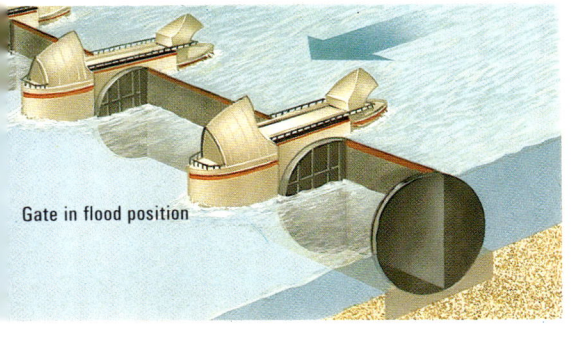

Above: Tongue Point, near Wellington, New Zealand. The flat tops of these cliffs were once a wave-worn beach washed by sea. When the coast rose, the beach was hoisted high and dry and a new beach formed below.

Marlborough Sounds in New Zealand. Here the sea has swamped V-shaped river valleys, creating a ragged coast with steep shores.

For centuries the people of the Netherlands have waged a continuous war against the sea. Much of their country is below sea level, and is protected from flooding by dykes. In this century the Dutch have reclaimed vast areas from the sea, including this land in the Province of Friesland now used for grazing cattle.

Ice at Sea

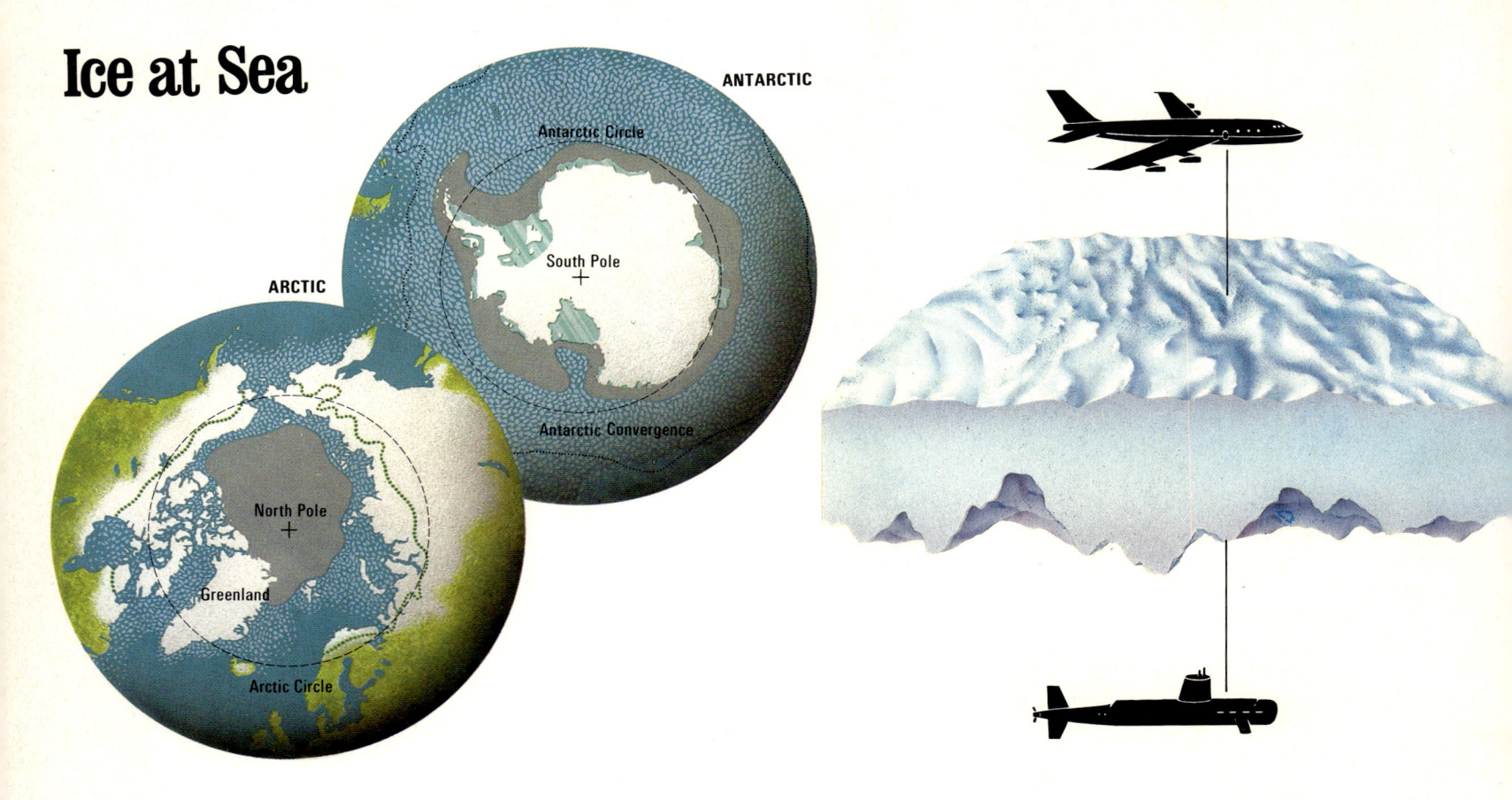

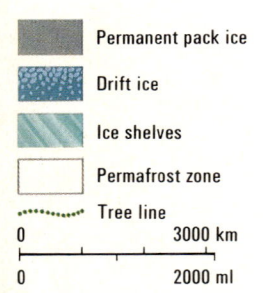

Above: Ice affects large tracts of the Arctic Ocean around the North Pole, and the Antarctic or Southern Ocean which surrounds Antarctica.

Legend:
- Permanent pack ice
- Drift ice
- Ice shelves
- Permafrost zone
- Tree line
- 0 — 3000 km
- 0 — 2000 ml

Above: A plane with a laser profilometer and a submarine with upward-looking sonar between them gauge the thickness of Arctic Ocean ice.

More than 1 in 10 parts of the world's total oceanic surface is covered by ice. Some of this is land ice which has flowed into the sea; some of it is frozen seawater.

Deep oceanic waters freeze less readily than shallow, inshore waters. As the surface water cools, it becomes denser and sinks, to be replaced by warmer water welling up from below.

Seawater freezes at about −1·9°C. First, a film of so-called frazil ice appears. Moving water sculpts this into thicker 'pancakes'. Later, slabs of pancake ice stick together to build still larger, thicker FLOES, that coalesce into a mighty sheet of PACK ICE. Wind pressure may open cracks called polynyas. Wind also forces immense floes to crash, driving one edge above another to create a huge pressure ridge, up to 58 metres (190 feet) thick. Snow falling on the frozen water builds up into thick compacted layers.

In summer, winds, currents and even tides break up much of the SEA ICE and carry it away as ice floes. The Antarctic ice pack is surrounded by open sea, and its ice floes spread out over 17 per cent of the world's ocean surface. As they drift, the floes melt and shrink.

The Arctic Ocean is virtually landlocked. Floes circulate around it, sometimes for centuries before escaping with the East Greenland current flowing out between Labrador and Greenland. Others gradually shrink and break up as they collide with each other.

Icebergs

ICEBERGS are chunks of land ice that have broken off into the sea. There are two main types. Tabular icebergs, mainly in the Antarctic, are formed when parts of the ICE SHELF – which extends from the land mass into the sea – break off and float away. They have flat tops with sheer sides up to 50 metres (160 feet) out of the water. They can be very large: one was spotted in 1956 which was bigger than Belgium.

There are some icebergs in the north which look like tabular bergs, but are much smaller. They probably come from the ice shelves of Ellesmere Island and Greenland.

The other kind of berg is castellated – so called because such bergs often resemble fantastic castles. They are formed when glaciers meet the sea and huge masses of ice break off and float away. Most of these bergs come from the glaciers along the coasts of Greenland and Antarctica. Greenland's glaciers alone may yield more than 14,000 icebergs a year. Some tower 90 metres (300 feet) above the sea, with another 360 metres (1200 feet) concealed below.

Like floes, icebergs drift according to the wind and currents. Some of the Antarctic bergs are so massive that they survive almost to the tropics.

Hazard to Shipping

Ice at sea can be a barrier and a danger to shipping. For centuries, sea ice stopped ships sailing around the north of America and Asia. In 1912, over 1500 passengers died when

Two kinds of iceberg. Above: A tabular iceberg typical of those that carve off from the ice shelves jutting from Antarctica. Right: A castellated iceberg, formed when a chunk of ice broke from a valley glacier where its front edge leaves the land and meets the sea.

the liner *Titanic* hit an iceberg in the North Atlantic.

But sea ice is not such a barrier as it once was. Now there are icebreakers that can plough through ice 2·1 metres (7 feet) thick, and nuclear submarines that can travel under the Arctic ice from the Atlantic to the Pacific Ocean. In the future, there may even be cargo submarines carrying goods from Europe and America to Japan via the Arctic Ocean.

Icebergs are still dangerous to shipping, and the INTERNATIONAL ICE PATROL exists to warn ships of icebergs drifting into North Atlantic shipping lanes.

One day, people may even drink the fresh water locked up in icebergs. Already there are plans for tugs to tow Antarctic bergs to thirsty California.

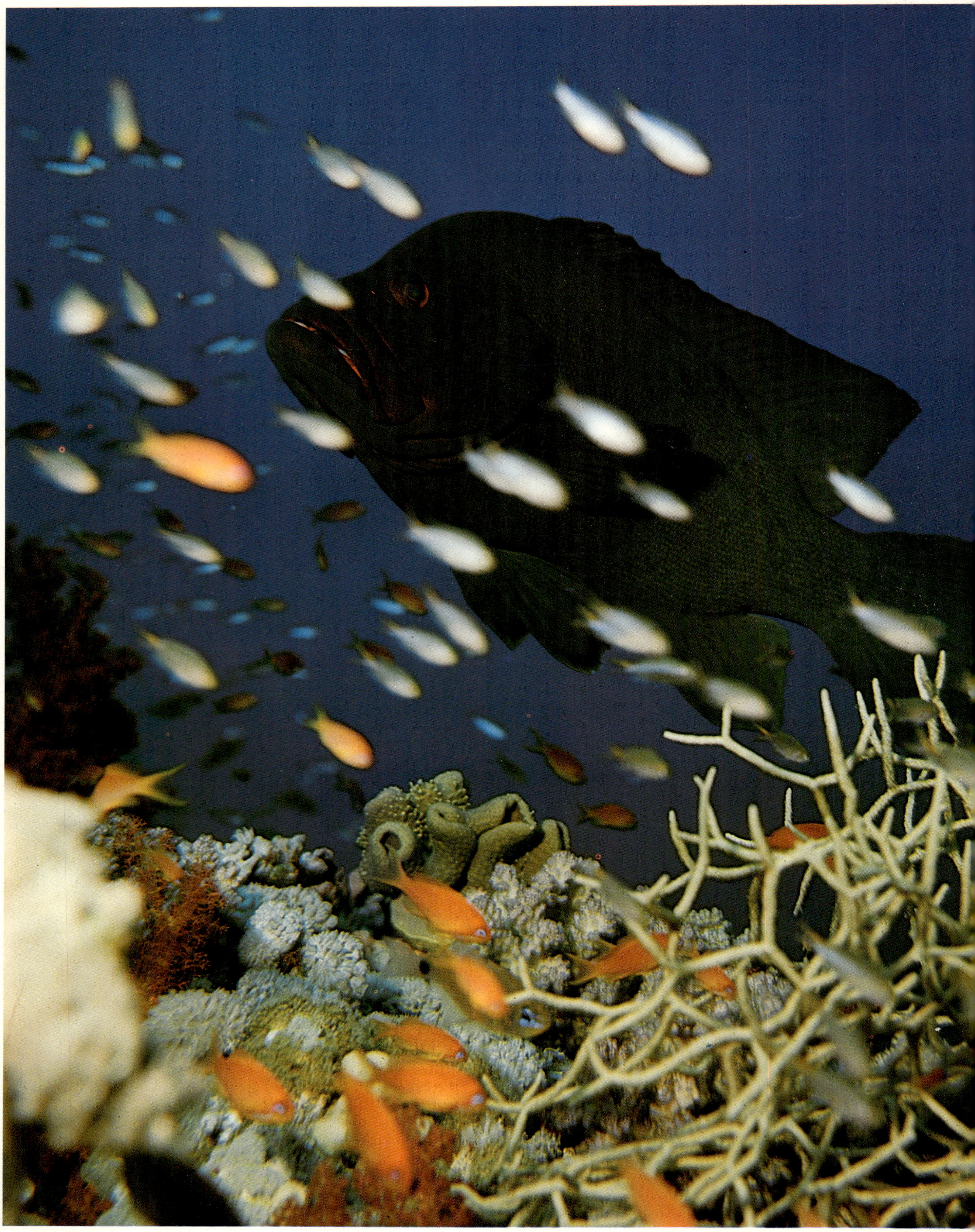

Chapter Three

LIFE IN THE WATERS

Life began in the ocean. Soft-bodied jellyfish, sponges, worms, sea cucumbers and seaweeds similar to those that flourished about 600 million years ago still survive in this stable habitat. Invertebrates related to sea cucumbers and starfish evolved into sharks and bony fishes. Some of these vertebrates developed into amphibians which in turn evolved into reptiles and mammals. Some of their descendants returned to the oceans; from them came today's whales, seals, sea cows and marine turtles – animals almost as much at home in the sea as fishes are.

There is life at every depth and latitude. Seaweeds and most animals thrive in the sunlit upper layers, but there are a few strange animals which manage to exist even in the blackness of the lower depths. And while shoals of exotically coloured fish live in the tropical coral reefs, other species endure polar oceans theoretically cold enough to freeze their blood.

People have long used the oceans as a source of food. But so many fish are now being taken from the sea that stocks of some species are being dangerously depleted.

A big predatory grouper lurks beyond small shoals of fish around a coral reef in the Red Sea. Predators are at the end of a complex food chain that starts with the minute plants of the plankton which first capture the energy of the Sun through the process of photosynthesis.

Left: A small crustacean larva and tiny floating algae. These scraps of life are members of the plankton – the myriads of drifting animals and plants that can transform the upper layer of the sea into a nutritious broth. Directly or indirectly this broth supports almost all the larger creatures living in the sea.

Right: The distribution of phytoplankton varies, as this map shows. The darker areas are rich in phytoplankton, in the lightest areas it is most scarce. The most prolific zones are those where top and bottom layers of the ocean mix. The abundance of plankton varies seasonally. Phytoplankton is most plentiful in the spring, zooplankton a few months later.

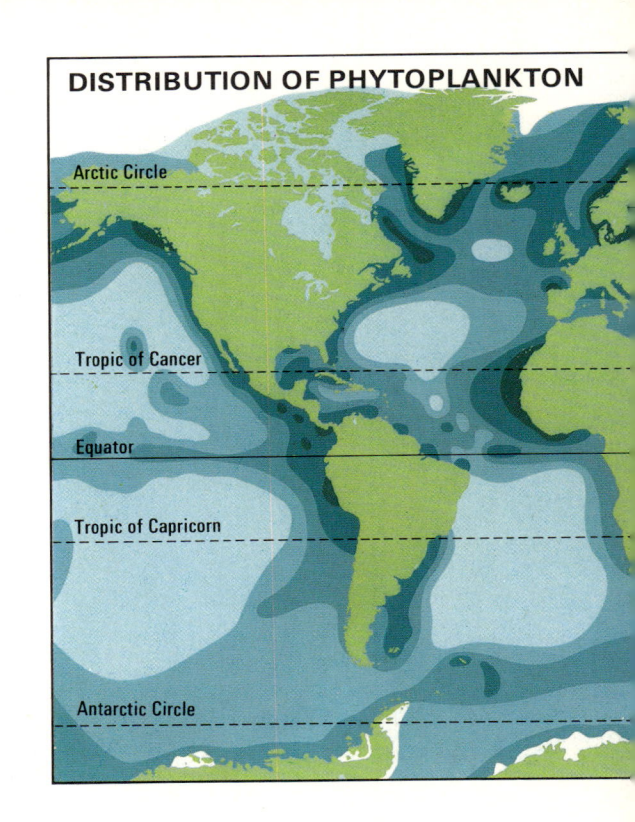

DISTRIBUTION OF PHYTOPLANKTON

Arctic Circle
Tropic of Cancer
Equator
Tropic of Capricorn
Antarctic Circle

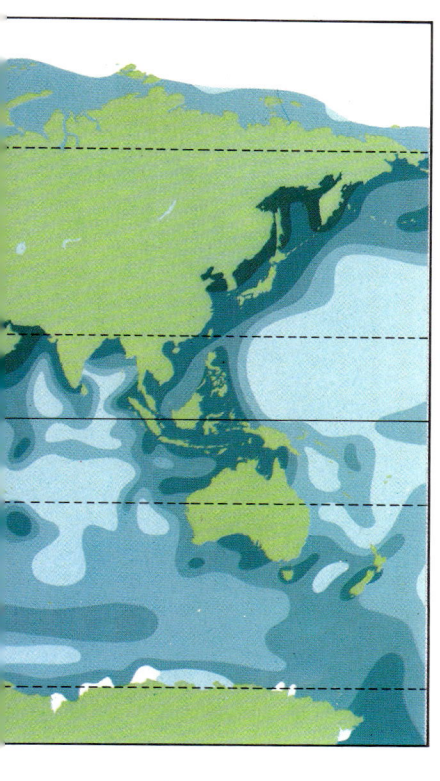

The Flow of Food

All animals ultimately depend upon plants for their food. Only plants can tap the energy in sunlight to manufacture food from water, carbon dioxide gas and inorganic salts in the presence of chlorophyll – a process known as photosynthesis. Sunlight cannot penetrate deep down into the sea, and so marine plants live only in the top 180 metres (600 feet).

Huge tracts of open ocean teem with PLANKTON: floating, drifting life found in the sea. Plant forms are called phytoplankton; animal forms are called zooplankton, and include baby sea worms, crabs, jellyfish, starfish and fish fry as well as much smaller organisms. While zooplankton feeds on phytoplankton, both kinds are eaten by a great variety of other, larger creatures. One form of zooplankton, KRILL, is the staple food of the huge baleen whales.

The inter-relationship between predator and prey throughout the seas is highly complex. For example, herrings feed mainly on COPEPODS, a tiny kind of shrimp, which itself feeds on phytoplankton. Herrings in their turn fall prey to cod. Phytoplankton, zooplankton, herring and cod are linked in one food chain. But herrings are also food for sharks, porpoises, gannets and people. There are many food chains, interweaving to form complicated FOOD WEBS.

At the same time, marine plants and animals form food pyramids. In a food pyramid vast numbers of tiny plants help to support just one top predator. For instance, 100,000 phytoplankton plants feed 10,000 zooplankton crustaceans, which nourish 1000 young herring – enough to satisfy only 10 codfish or just one porpoise.

Many animals and bacteria on the seabed survive on the decaying remains of living things. By breaking down these remains, they release minerals and nutrients into the sea for plants to use again. Thus sea life also forms a great food cycle.

Below: Sea creatures and plants living at different depths. Depth bands and sizes of organisms are not shown to scale, and cool-water and warm-water organisms are mixed. Most life thrives in the sunwarmed, sunlit surface. There, phytoplankton 'meadows' are browsed on by zooplankton. Plankton directly or indirectly feeds fishes, invertebrates, seabirds and whales. As surface organisms die and sink, they yield food for strange creatures of the lower depths. Here, life is scarce but varied. Dwellers of the deep ocean bed include burrowers, and fishes whose stilt-like fins prop up their bodies and allow them to rest on soft ooze without sinking. Despite zonation by depth, many creatures nightly commute 400 metres (1280 feet) upwards to feed.

Animal groups in the sea	
Members of all 22 major animal phyla (divisions) live in the sea, compared with only 9 on the land. But its greater variety of habitats means that the land has 85% of all species. Below is a list of phyla and their locations.	
L = Land	
F = Freshwater	
S = Sea	
Protozoa: protozoans	LFS
Porifera: sponges	FS
Coelenterata: jellyfish, corals, hydroids	FS
Ctenophora: comb jellies	FS
Platyhelminthes: flatworms	LFS
Nemertina: ribbon worms	LFS
Rotifera: wheel animalcules	LFS
Nematoda: nematode worms	LFS
Priapulida: 'worms'	S
Sipunculoidea: 'worms'	S
Echiuroidea: 'worms'	S
Pogonophora: 'worms'	S
Annelida: segmented worms, eg fanworms	LFS
Arthropoda: includes crustaceans, eg crabs	LFS
Mollusca: molluscs, eg mussels, squids, snails	LFS
Phoronida: 'worms'	S
Bryozoa: bryozoans	FS
Brachiopoda: lamp shells	S
Chaetognatha: arrow worms	S
Echinodermata: echinoderms, eg starfish	S
Hemichordata: hemichordates	S
Chordata: chordates	LFS

The Undersea World

Countless animals breathe, move, feed and breed beneath the waves. They do so in many different ways, and come in many shapes and sizes. They range from the streamlined fishes and elongated worms, to plant-like sponges, sea anemones and sea squirts, spiny sea urchins, and shell-protected crabs and sea snails.

Soft-bodied creatures like JELLYFISH and sea anemones need the buoyancy of water to support them; while vertebrates like the giant whale would suffocate under their own weight without the support of the sea.

Since they are exposed to air, most land animals need a waterproof skin to prevent their moist insides being dried up. Being perpetually bathed in fluid, sea creatures need no such protection. Jellyfish, sponges, sea slugs, sea squirts and sea worms have damp, soft skins; fishes' scales also have a mucous outer coating.

Unlike land animals, most marine creatures respire oxygen dissolved in water. Fishes take in water through their mouths and push it out through their GILLS which extract oxygen from it. MOLLUSCS such as clams and whelks, and CRUSTACEANS such as crabs, shrimps and lobsters, also breathe through gills. Some tubeworms have gills which sprout from the head like tiny palm fronds. Sponges, HYDROIDS, sea mats and sea squirts draw in oxygen-bearing water by means of tiny hairlike CILIA.

Moving in Water

Sea animals also move in different ways. Most molluscs 'walk' by means of a fleshy foot. But scallops and squid can jet forward by squirting water backward. Starfish and sea urchins progress on myriads of tiny so-called tube feet. Paddleworms swim with paired paddles jutting from their body segments. Crustaceans have a variety of limbs designed for swimming, leaping or walking. Among the plankton are microscopic creatures that swim by lashing water with cilia.

But no animals swim more gracefully and efficiently than the fishes. These vertebrates form two main groups: the BONY FISHES, and the CARTILAGINOUS FISHES – sharks, rays, and

Left: Some different ways of moving in water. Fish are streamlined and can use their tails to move rapidly through the water, and their fins to balance and steer themselves. Jellyfish and squid move forwards by pushing the water out behind them. Crabs walk sideways on tiptoe along the seabed.

Far right: Only the armoured forepart of this hermit crab projects from the empty whelk shell in which the crustacean has tucked the soft, coiled, vulnerable hind part of its body. When the hermit crab outgrows one shell, it simply slips into another. For many a hermit crab, the stings of sea anemones living on its adopted home give added protection from predators.

Below: Purple and orange sponges in the Caribbean Sea. Sponges derive their scientific name *Porifera* ('pore-bearers') from the tiny pores covering their bodies. Food and oxygen enter a sponge through pores and travel through canals to chambers which are drained by more canals that empty into a large opening in the body.

chimaeras (the latter commonly called the rat-tail fish). Both groups include fast, streamlined swimmers. Most fishes have a swim bladder to help them adjust to different pressures at different levels in the water. Sharks, however, do not have a swim bladder, and must keep swimming or sink.

Hunters and Hunted

Between them, marine creatures exploit all the food sources available. Sponges, sea anemones and hydroids, and also many burrowing BIVALVES and worms, stay put and wait for food to come to them. Devices that act as nets, poisoned darts, lassoes or vaccuum cleaners trap the foods they need. These foods range from plankton and scraps of DETRITUS sucked in by razor shells to quite large fish caught by the tentacles of an octopus that lies in ambush.

Fish are mainly active hunters, and some are formidably armed with teeth. But many kinds are specialized for one main way of feeding. Whale sharks – the largest fish of all – browse on plankton and small fishes. Swordfish slash out with their sword-like upper jaws to stun smaller fishes that swim in shoals. White sharks may savage other sharks. Demersal (bottom-dwelling) fish such as skate, dogfish and halibut tend to have their mouths under their heads, well placed to snap up seabed worms and shellfish.

Against the hunters is ranged a strong array of defences. Slow-movers such as crabs and lobsters rely largely on body armour for protection. Plaice and turbot change colour to match the seabed below them. Dark upper parts and silvery bellies make surface-swimming mackerel and tuna hard to see from above or below. Fishes can also swim fast if they are pursued. Herring, mackerel and other fish that swim in shoals gain protection by presenting a large and confusing target.

Breeding in Water

The eggs of most sea creatures are fertilized outside their bodies, with water as the medium to bring the male sperm and the female egg together. Fish eggs and young fish provide food for many other animals: thousands of eggs are shed into the sea for each one that is successfully fertilized and survives to maturity.

Above: A bony fish breathes in water through its mouth and exhales it through gill slits at the sides of the head. Each side has four gill arches equipped with soft filaments. These extract four-fifths of the oxygen dissolved in the water passing through. On each side of the head, used water then enters a single cavity, protected from the outside by a bony plate. As this gill cover opens, the water escapes.

Above: A sea anemone with tentacles extended. Stinging cells in the tentacles capture small passing creatures with tiny poisoned threads. These paralyse them. Then the tentacles pull their prey down into the mouth located at the top of the sea anemone's tube-shaped body.

Right: Sharks are swift, streamlined hunters. Overlapping rows of sharply pointed teeth help big killers like the great white shark to trap fish or sever limbs from human swimmers. But many sharks pose no threat to man. Some have broad, flat teeth for crunching molluscs. The whale shark – at up to 18 metres (59 feet) the largest fish of all – eats only plankton.

Sea Mammals

Mammals evolved to breathe atmospheric air, and to feed, move and breed on land. Yet many millions of years ago some began returning to water, the element from which their ancestors the lobe-finned fish had emerged.

This move produced sea otters, sea cows, seals and whales – creatures almost as well adapted to an ocean life as the fishes some so much resemble. Seals and some sea otters come ashore to breed. The rest feed and give birth under water. But all sea mammals must surface to breathe air.

Sea Otters and Seals

Sea otters live offshore around the North Pacific Ocean. They swim with hindlimbs shaped like paddles, diving to the seabed for shellfish and stones. Floating on its back, a sea otter lays a stone on its chest and bangs the shellfish on this 'anvil' until the shell breaks open. Sea otters that live off California feed and breed at sea. But those off northeast Asia bask, and bear their cubs or 'kits' on reefs.

All four limbs of the seals and walruses have evolved as flippers. On land the eared seals (sea-lions and fur seals) can bring their hindlimbs forward, lift their bodies off the ground and actually gallop. Walruses can do this too. But the hindlimbs of true seals always trail. These seals move clumsily on land, but all seals swim splendidly. Dense hair or thick blubber (fat) beneath the skin insulates their bodies from the cold.

Like other mammals, sea mammals suckle their young. Thick blubber insulates these mammals from polar cold. In summer these walruses keep cool by radiating heat from blood vessels near the body surface. The blood that flows through the vessels makes their entire bodies blush.

Walrus 3·5 metres (10–12 feet)

Grey seal 2·4 metres (8 feet)

Manatee 1·8 metres (6 feet)

Sea lion 1·8 metres (6 feet)

Sea otter 0·9 metres (3 feet)

Statue of the Little Mermaid in Copenhagen harbour. Sailors long ago used to report that they had seen these mythical creatures sitting on rocks combing their hair. What they actually saw were probably manatees.

Seals hunt under water for fish, shellfish or crustaceans. (Walruses prise clams and mussels from the rocks with their tusks.) Most seals live in cool, fish-rich waters of high latitudes. Most species go onto the land only to bask, breed or moult.

Sea Cows and Whales

Manatees and the dugong (the two sea cow families) browse sluggishly on underwater plants off low tropical coasts.

No mammals are better built for living in water than WHALES. Their streamlined, blubber-insulated bodies taper from the head to the tail which consists of two horizontal flukes. The forelimbs are broad flippers. Most whales breathe in and out through a blow-hole or two slits on top of the head. Some can remain submerged for up to two hours before they rise to breathe again.

Whales make up two great groups: the baleen and the toothed whales. Swimming open-mouthed, baleen whales trap millions of tiny planktonic krill on a sieve composed of baleen plates that hang from the mouth roof like the frayed teeth of a giant comb. Baleen whales include Earth's largest animal: the blue whale. One slaughtered specimen measured 29·48 metres (96¾ feet) and weighed about 177 tonnes (174 tons).

Toothed whales, such as the dolphins, porpoises and killer whales, are much smaller than baleen whales, but feed on larger prey. Most eat fish, but killer whales kill penguins, seals, and even other whales.

Whale calves are born under the water. Unlike other mammals, they are born tail first. They can swim as soon as they are born, and so are able to rise to the surface to breathe air.

Above: Dolphins in the Atlantic Ocean. All sea mammals have to surface to breathe air, yet one breath is enough to allow some whales to stay submerged for up to two hours.

Below: Members of the mammal groups evolved for life in the sea. Most kinds would be secure except for man. Hunters killing for meat, oil, fur or ivory have decimated some species, killed off one kind of seacow and almost annihilated the blue whale.

Blue whale
24 metres (80 feet)

Sperm whale
18 metres (60 feet)

Killer whale
12 metres (40 feet)

39

Life in the Depths

Left: Black angler fish. The light from its lure comes from bacteria that glow in the dark. This process uses oxygen, evidently provided by the fish's blood supply. Scientists think that an angler fish controls the brightness of the light by varying the amount of blood that reaches the tip of its lure.
Centre: *Astronesthes* – a black, scaleless fish with long fangs and a voracious appetite. Its body shines with many light-producing organs. (Most deep-sea fish glow when oxygen and luciferin combine in special skin glands; many control light to scare, attract or show up other animals.)
Right: Its long, tapering tail helps the rat-tail to tilt its body as it grubs for starfish and shellfish on the seabed.
Below: *Eutaemophorous*, a rare deep-sea fish. This individual's stomach bulges after a feast of copepods, members of the zooplankton that spend daylight hours deep down.

In contrast to the sunlit upper layers of the ocean, the lower depths are dark. Here no green plants grow to provide food for the inhabitants. Squid, prawns and LANTERN FISH rise at night to feed on zooplankton. Other deep-sea animals prey on these small commuters or on one another. A third group feeds upon the rain of dead and dying plants and animals that sinks down from above. But there are far fewer animals living in the deep sea than in the surface layer or EUPHOTIC ZONE.

The deep-sea creatures have adapted to the everlasting dark in many unusual ways. Most fishes, for example, have big eyes to help them detect what light there is. And because food is scarce, most species grow no larger than a person's hand.

The twilight zone begins about 180 metres (600 feet) down. The deeper you go the stranger the fishes become. At 250 metres (820 feet) down, fishes have rows of light-producing organs and silvery sides like mirrors. Lights and reflective body surfaces help the fish to see and also help to disguise their shape from predators. At 700 metres (2300 feet) lurk dark, drab fishes, and scarlet prawns. Even these prawns look black because red light cannot penetrate deep enough to reveal their colour. But no doubt the vibrations set up in the water by their movements betray them to their hunters.

Fishes equipped with vast mouths and elastic stomachs make the most of the rare meals that come their way.

Below 1000 metres (3280 feet), food is too scarce for fishes to waste much energy pursuing it. Instead, beasts like the angler fish have luminous lures that help draw prey to them.

The deep-sea zone below 3700 metres (12,000 feet) is known as the ABYSS. Life is sparse down there, but it persists even at the greatest depths. Below 10,000 metres (32,800 feet), animals act as scavengers and include sea cucumbers (which are related to sea urchins and starfish), worms, crustaceans and lamp shells. They suck in nourishment from animal remains or droppings which have drifted down to the mud.

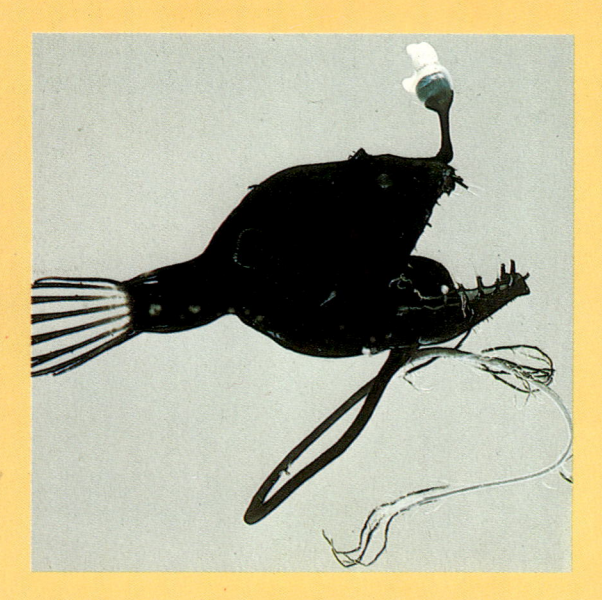

Polar Regions

The cold waters of the Arctic and the Antarctic regions have lifestyles of their own. But here, as in warmer seas, all life depends upon surface-dwelling phytoplankton. This grows profusely in Antarctic waters, where upwelling currents bring nutrient salts to the surface. In the Arctic Ocean, water layers mix less readily and plankton tends to be less plentiful. But in both regions, a springtime BLOOM of phytoplankton triggers an explosive 'flush' of zooplankton.

Some tracts of sea hold millions of the shrimplike crustaceans called krill – the staple diet of baleen (whalebone) whales, squid, and certain seals and seabirds.

Adapting to the Cold

Many organisms of the polar seas tolerate cold or cannot live in warm conditions. The Antarctic krill *Euphausia superba*, for example, dies in water above 4°C (39°F) and cannot thrive north of the boundary called the Antarctic Convergence.

So-called cryopelagic fishes spend part of their lives close to sea ice. The polar cod, for instance, haunts crevices in ice floes where it eats zooplankton and hides from its many enemies including seabirds, narwhals, white (beluga) whales, and the bearded, harp and ringed seals.

Astonishingly, polar fishes swim happily in water colder than the freezing point of blood. Natural antifreezes in their bodies help to make this possible.

Between them, fur, feathers and blubber insulate polar bears, penguins, seals and whales against the polar cold. All of these animals find food in polar seas and some never set foot on dry land: whales because they cannot leave the water; emperor penguins and harp seals because they raise their young on sea ice.

Above: Antarctic cod, a big-headed, bottom-dwelling fish, found in Antarctic waters. In winter, the blood of such fishes gains increased concentrations of chlorine, potassium and sodium ions and urea. This lowers the temperature at which the fishes' blood will freeze, much as salting a wet road in cold weather stops ice forming.

Above: All the animals on this ice shelf jutting from Antarctica depend upon the ocean for their food. The crab-eater seal and the small Adélie penguins catch shrimplike krill. Emperor penguins hunt squid. Both kinds of penguin also feed on fish. Predatory skuas and scavenging sheathbills get food largely indirectly from the sea. Skuas steal penguin chicks and eggs. Sheathbills eat carrion and shoreline seaweed.

Right: Polar seas at least in part support every creature in this springtime Arctic scene. Ringed seals eat krill and polar cod. Polar bears hunt ringed seals. In winter, Arctic foxes trek far out on sea ice to scavenge at the kills of polar bears. King eider ducks, little auks and razorbills all feed in chilly northern waters. When winter ice locks up their Arctic Ocean larder, the birds must find food farther south.

41

HOW A CORAL REEF FORMS

Coral growing in warm and shallow water has formed a reef around an island created by volcanic action.

The sea rises or the island sinks, but coral continues growing upward.

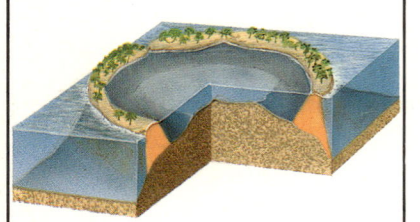

The island has vanished, but coral reefs remain, to form an atoll.

Tropical Seas

The warm waters of the tropics generally support more kinds of animals than polar oceans, but fewer individuals of each kind. Life is less prolific here because the warm surface layers of the ocean float above the cold, dense, deeper waters with little mixing. Thus nutrients that sink down from the upper levels are not readily replaced.

A 'Desert' in the Sea

Some ocean areas are particularly poor in plankton, and the animals which depend on it for their food. One famous ocean 'desert' is the Sargasso Sea, a still region circled by the major currents of the North Atlantic. Pushed in by certain currents, a thick floating layer of brown seaweed covers much of the Sargasso Sea. This sargassum weed helps to curb phytoplankton growth by limiting the light that penetrates the ocean. But the weed provides a foothold for hydroids, sea anemones and barnacles, and a hiding place for the strangely shaped sargassum fish.

Coral Reefs

Coral REEFS are richer in life. They include offshore FRINGING REEFS; BARRIER REEFS cut off from land by a wide channel; and oceanic ATOLLS, enclosing a lagoon.

The rock that is commonly called coral is a hard, limy substance formed only under water. The wave-battered, ocean-facing flanks of coral reefs are often substantially formed of calcium carbonate deposited by certain seaweeds. But the major coral builders are tiny relatives of sea anemones.

Each of these coral POLYPS builds and lives in a protective limy cup. At night the creatures thrust out tentacles to trap the minute drifting organisms that form their diet. When coral polyps die, their homes become a base for other polyps to build upon. Where millions of polyps form a colony, the result is a vast, upward-growing coral mass. Some CORALS look like brains, others like fans and stags' horns.

But corals do not grow below 45 metres (150 feet). This is because they rely on tiny plants called ZOOXANTHELLAE, which live inside their bodies and only thrive in well-lit water. Also, coral polyps cannot live where the temperature of the water drops below 18°C (64°F). For these reasons, coral reefs occur in the tropics, in shallow water away from muddy estuaries.

Life on the Reef

Many different kinds of animal live in coral reefs. Butterfly fish with slender snouts probe the coral heads for molluscs and crustaceans. Clownfish, damselfish and others flaunt their colourful bodies to keep rivals away from their territories. Sponges, tubeworms and sea squirts live on old coral heads and suck in tiny drifting organisms. Crevices hide scavenging crabs, sea urchins, and predatory moray eels.

As well as using the coral reef as a home, some animals eat the coral polyps themselves. Parrot fish bite off and crunch up chunks of living coral. The crown-of-thorns starfish swarms over coral heads, dissolving and digesting polyps. This particular predator has badly damaged parts of Australia's Great Barrier Reef – the largest reef on Earth. Then, too, date mussels, sea snails and palolo worms all burrow into dead coral, helping to undermine and break up reefs.

Right: A clownfish among the normally lethal tentacles of a sea anemone. This fish can use the anemone as a safe hiding place. The sea anemone profits from food scraps that the clownfish drop.

An angel fish among coral in the Caribbean. Hard coral builds up into reefs which provide a home for many species of colourful fish. Some of the fish eat parts of the coral.

The strangely shaped sargassum fish hides among sargassum weed floating in the windless Sargasso Sea. Knobs and tassels projecting from the fish resemble the weed's own tassels and air bladders.

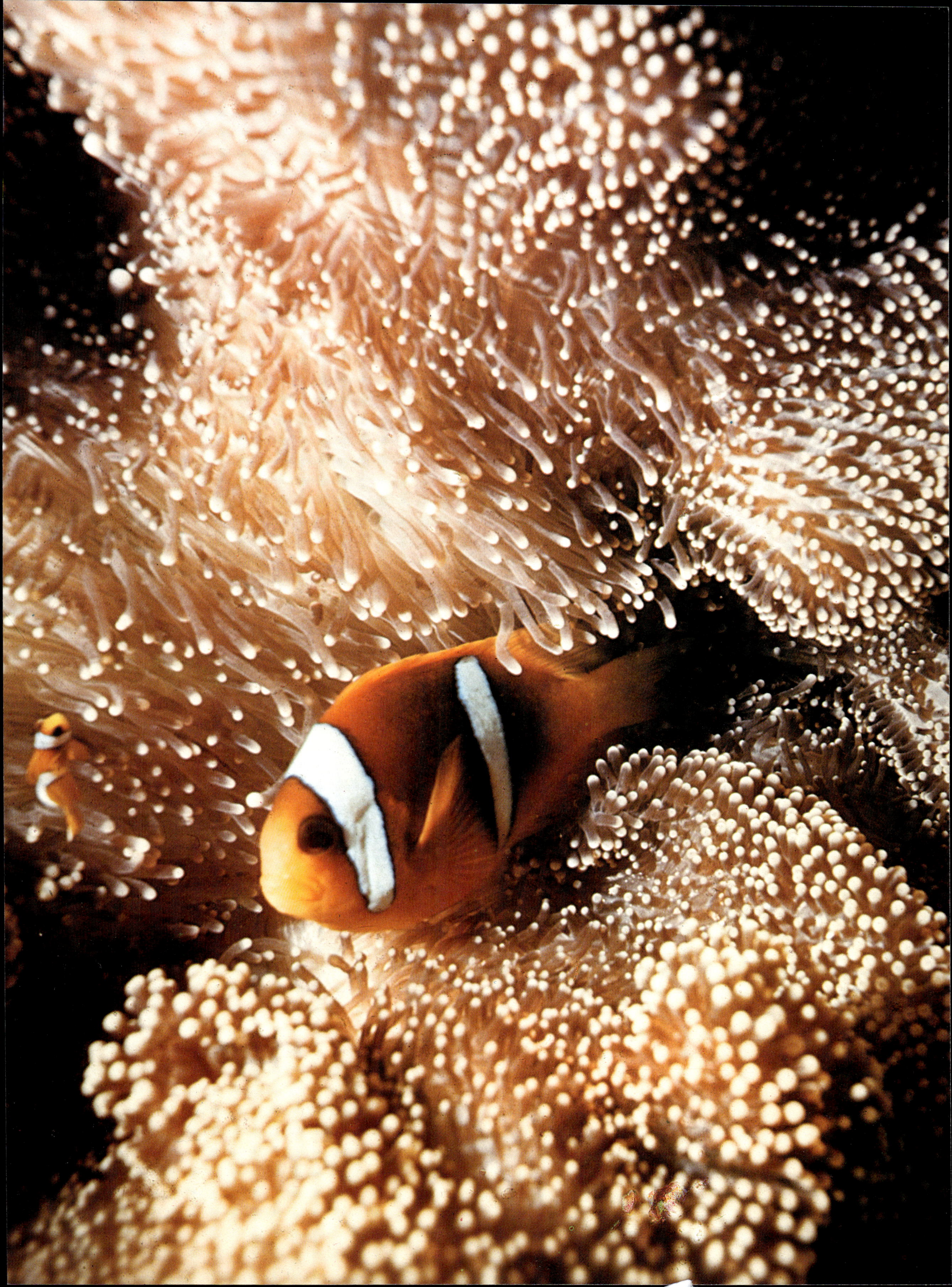

Fishing

Back in Stone Age times, people used to gather shellfish at the sea's edge – a very basic form of 'fishing'. Today, millions of people, particularly in tropical and sub-tropical regions, still use simple equipment to catch fish. Some nations, however, have developed large commercial fishing fleets.

Fish contain useful proteins, fats, minerals and vitamins, and are valuable food for humans and livestock. The annual fish catch worldwide now stands at about 70 million tonnes. The largest hauls come from shallow seas and over the continental shelves.

Some of the most fertile areas are those where upwellings occur: cold water rises, bringing with it essential nutrients that enrich surface water where plankton grows. This is the case in the North Atlantic and North Pacific, and off Namibia, California and Peru. Tidal currents also help to stir up the water in the shallow areas off the rims of continents. Banks – elevations of the sea-floor – can also produce upwellings and fertile fishing grounds. (See the map of the distribution of plankton, page 34.)

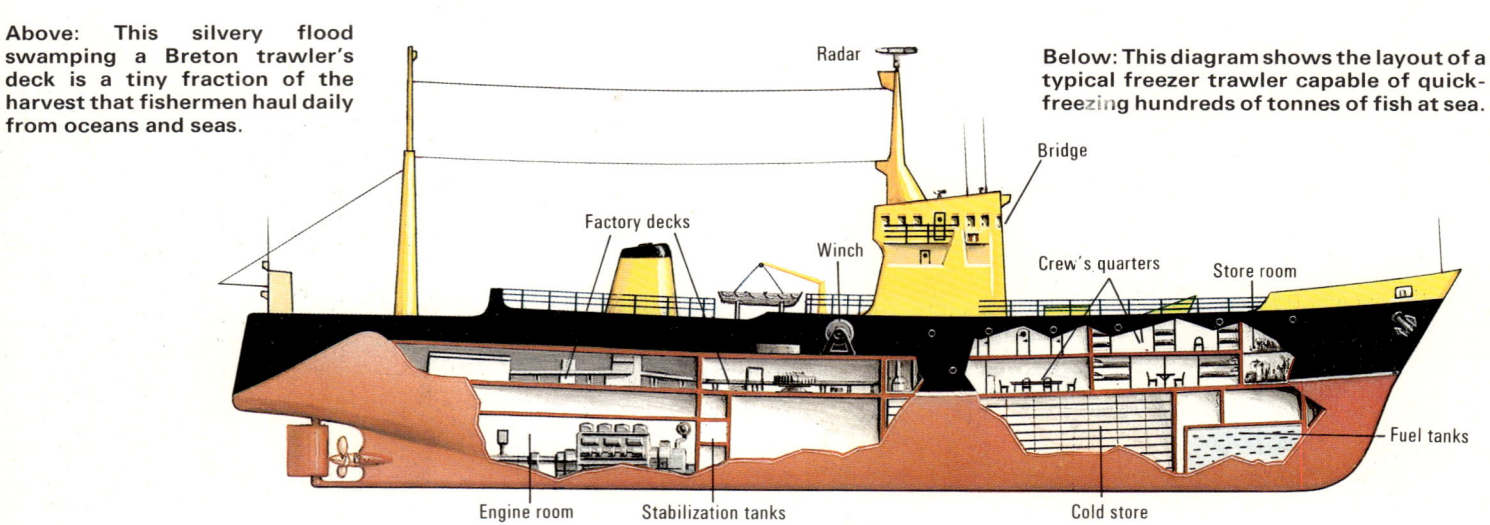

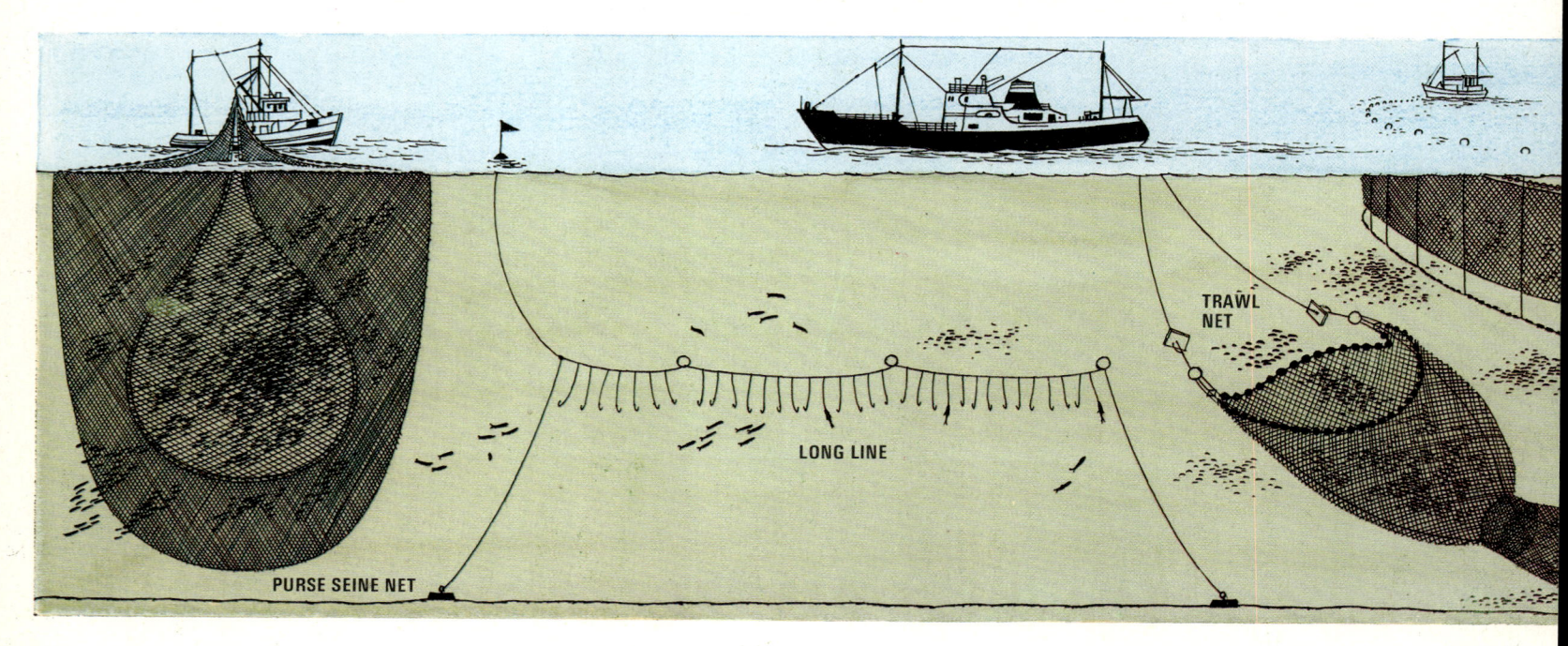

Above: This silvery flood swamping a Breton trawler's deck is a tiny fraction of the harvest that fishermen haul daily from oceans and seas.

Below: This diagram shows the layout of a typical freezer trawler capable of quick-freezing hundreds of tonnes of fish at sea.

44

Various fish are caught at different depths. Pelagic fishes like sardines, mackerel and herring tend to swim near the surface. Demersal roundfish, such as cod, and flatfish, such as plaice, are hunted on the seabed.

Different species are commercially important in different oceans. In the North Pacific, for example, those bottom-dwelling fishes Alaskan pollack, cod, flounder and rockfish figure heavily. The Pacific also yields three kinds of tuna – a big, meaty, surface swimmer related to the mackerel. Tuna is also taken from the Atlantic and the Indian Oceans. Cod, haddock, saithe, whiting, hake and other demersal members of the order Gadiformes (codfishes) dominate North Atlantic fisheries.

Modern Fishing Methods

Nowadays, fishing fleets can use equipment that makes fishing less open to chance. Fishery laboratories are able to detect fish spawning grounds and migration routes, and sonar helps fishing vessels to home in on their prey. Sound impulses emitted by a forward-scan sonar device and echoed back by fish can reveal a shoal 4·8 kilometres (3 miles) away. Sonars are also used to guide the fishermen in actually catching the fish.

The illustration below shows some modern methods of catching fish in bulk.

Disappearing Shoals

Modern long-range fishing fleets are geared for reaping massive harvests. Russia and Japan particularly scour the seas of the world with small catcher craft that supply fish direct to big factory ships. The fish are then processed and frozen on board these ships while they are still at sea.

Below: Long-lining catches fish that do not form tight shoals. Trawl nets trap bottom-dwelling fish. Otter boards hold the net mouths open. Purse-seining involves pursing (closing) a net curtain around surface-swimming shoals. In drift-netting, 30 catching vessels may serve one mother ship, shooting their nets after midday and pulling them in after midnight.

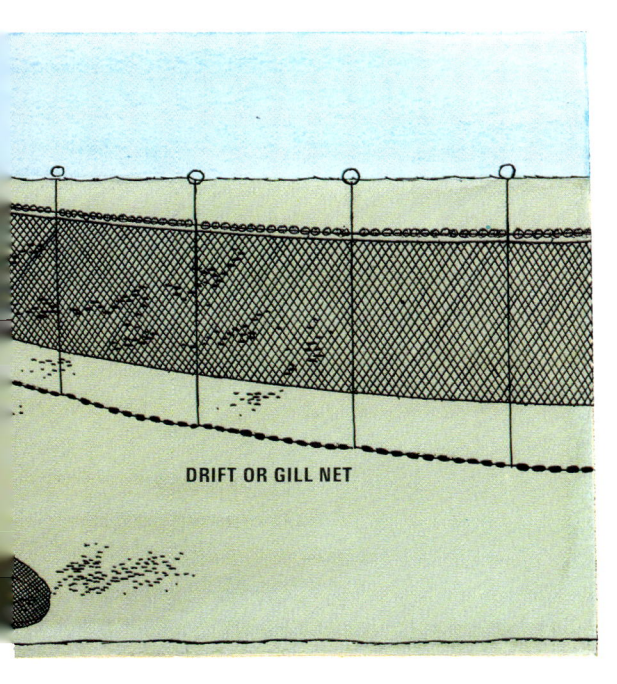

DRIFT OR GILL NET

But intensive fishing is now contributing to shortages of, for instance, herring, cod, Peruvian anchovies, and certain whales. To defend stocks hit by foreign fishing fleets, some coastal nations claim fishing rights extending up to 320 kilometres (200 miles) out to sea. Within these limits, other nations may be set strict quotas, or forbidden to fish at all.

Such measures may not halt the world's drain on fish stocks. If so, we shall have to turn to unexploited species, such as the grenadier, the director and the black scabbard. We are already beginning to *farm* fish as well as hunting them. We may yet have to apply to seafood the principle now being suggested for food produced on land: just as we could eat grain directly rather than the animals that feed on it, so we may one day have to bypass fish altogether and feed on plankton such as krill.

Above: A British fish farm consisting of a complex sea cage for rearing turbot. As stocks of wild-caught fish shrink, biologists explore schemes to farm fish like sheep or cattle. They have raised fish in tanks, salt-water ponds, and cages in the sea. But marine invertebrates and plants may do more than fish to meet rising global need for food.

World commercial marine catch in 1976	thousands of tonnes
Herring, sardines, anchovies, etc	15,089
Cod, haddock, etc	12,115
Miscellaneous	8350
Jacks, mullets, etc	7389
Redfish, bass, etc	4950
Molluscs	3851
Mackerel, snoeks, etc	3340
Tuna, bonito, etc	2209
Crustaceans	2023
Seaweeds	1189
Flounders, halibut, sole, etc	1123
Shads, milkfish, etc	697
Salmon, trout, etc	555
Sharks, rays, etc	533
TOTAL	63,413

Left: A Russian factory ship being supplied with fish by a small trawler. Factory ships of up to 40,000 tonnes may carry 14–50-tonne catcher craft on their decks to fishing grounds half way around the world. In one four-month stint, such fishing fleets can process 10 million cans of fish; 10,000 tonnes of fish products; 1000 tonnes of fishmeal, and 100 tonnes of fish oil.

46

Chapter Four

MAN BENEATH THE SEA

The human skeleton supports soft, moist tissues that are bathed with salty fluids and are packaged in a skin. Thus far we resemble the sea beasts from which we evolved. But humans have lungs which are built for breathing warm atmospheric air, and limbs which are adapted for moving under atmospheric pressure. The ocean – mankind's ancestral home – is hostile to the human body. Even strong swimmers grow chilled and drown if they stay too long in the sea.

So the undersea regions remain Earth's last great unexplored area. To investigate the submarine world of creatures, wrecks and minerals, people are now breaking through the water barrier, with aids for breathing, moving and keeping warm beneath the waves. Divers' suits, submersibles, and underwater habitats are helping explorers and technologists to open up the shallow offshore waters – a tract of sea equivalent in total area to Asia, the largest continent on Earth.

Resembling some strange being from an alien planet, a man prepares to film an underwater coral 'garden'. Coldproof, waterproof clothing, breathing and swimming aids, and special tools enable divers and submariners to work and even live for days below the waves where normally a human being can survive for only minutes.

BREATHING UNDER WATER

Deprived of oxygen, the brain dies in minutes. These pictures show the major ways in which divers under water get the oxygen they need.

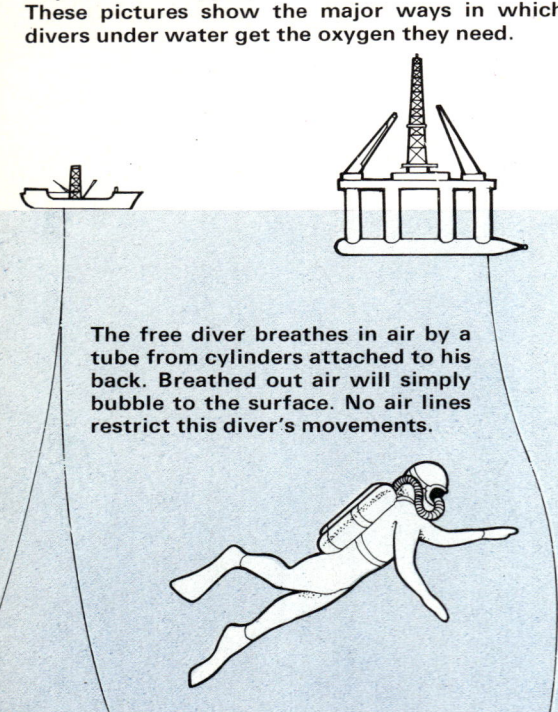

The free diver breathes in air by a tube from cylinders attached to his back. Breathed out air will simply bubble to the surface. No air lines restrict this diver's movements.

This diver breathes in air pumped down from the surface by means of a hose. Each time the diver inhales, a valve in his mouthpiece opens and lets fresh air inside his helmet.

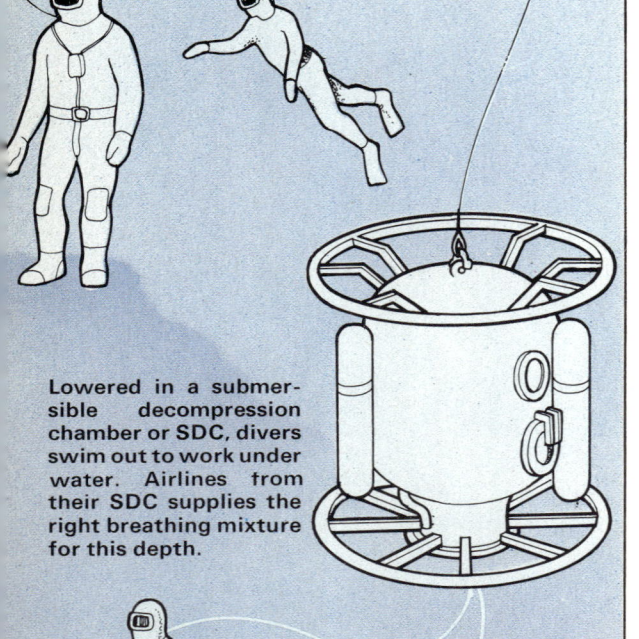

Lowered in a submersible decompression chamber or SDC, divers swim out to work under water. Airlines from their SDC supplies the right breathing mixture for this depth.

A breathing mixture with helium instead of nitrogen proves best for deep diving. Unlike nitrogen, helium at high pressure does not stupefy. But in rapid compression it causes trembling, dizziness and pain. Adding a trace of nitrogen stops this.

Divers

Early Divers

For many centuries people have been diving down into the sea to hunt for pearls and sponges. Early divers plunged in naked, and simply held their breath. They could not dive very deep or stay down for very long.

But since the early 1700s, inventors have created a variety of aids to help people to swim or walk about, at least in shallow waters.

The first breakthrough was the waterproof DIVING SUIT developed in the early 1800s. Air pumped from the surface passed through a tube into a heavy metal diving helmet with glass windows. At first, breathed and surplus air just bubbled out from an opening beneath the helmet, and the diver risked drowning if he bent his head. Improving on this early model, Augustus Siebe in 1830 added a valve that let air out without admitting water.

Soon, divers wearing helmets and weighed down by lumps of lead did valuable salvage work on sunken wrecks. But many divers fell mysteriously ill or died. We now know why. The deeper a diver descends, the greater the pressure of water acting upon his body. He must breathe air or gas at a similar pressure, but this causes problems.

DIVER'S CLOTHING

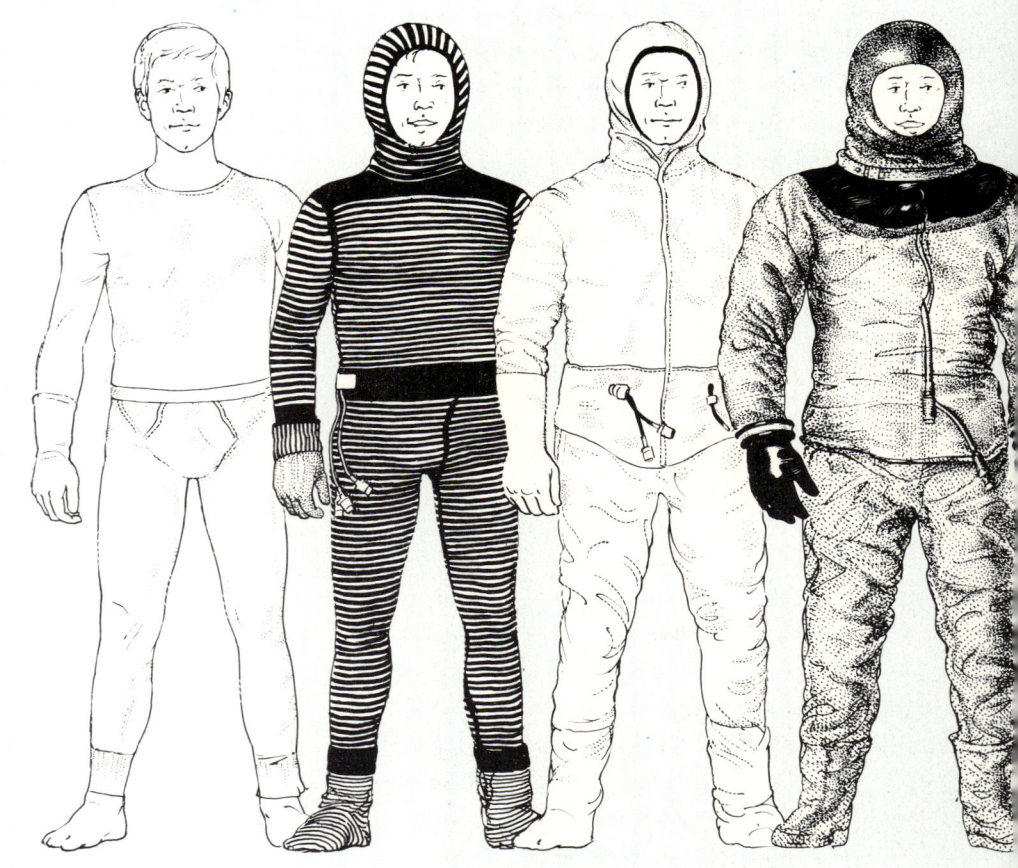

1 Cotton undergarment
2 Electrically heated woollen suit with insulating filaments knitted into it
3 Nylon-pile insulating garment
4 Dry suit

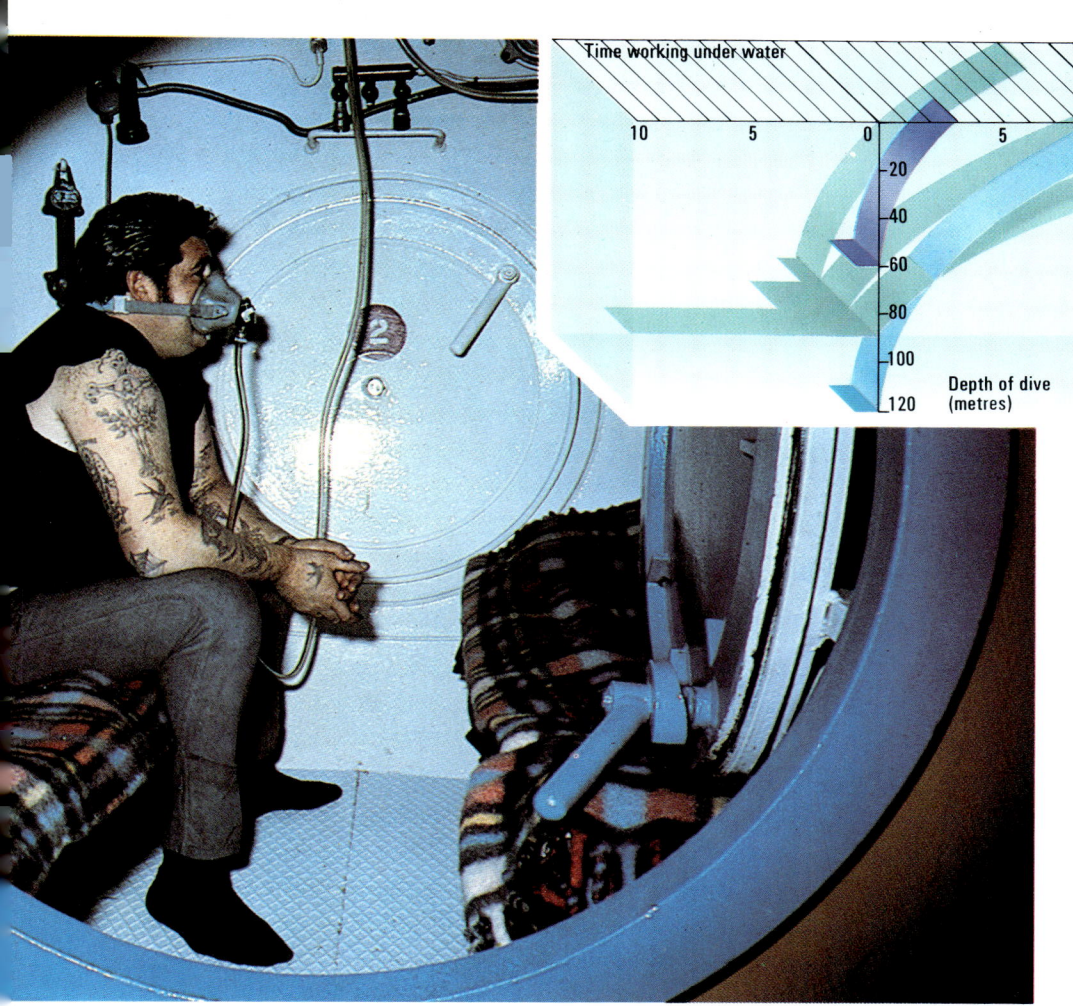

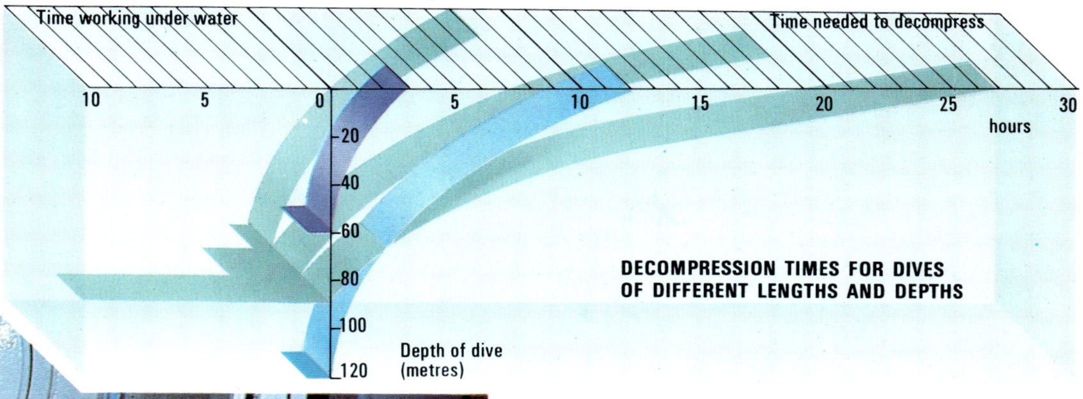

DECOMPRESSION TIMES FOR DIVES OF DIFFERENT LENGTHS AND DEPTHS

Left: A diver in a decompression chamber aboard an oil rig. After a long, deep dive, he is placed here in a breathing mixture at the same high pressure as the water where he worked down in the sea. As the hours pass the pressure is reduced until it matches that of the sea-level air outside the chamber. The diver may then step out safely.

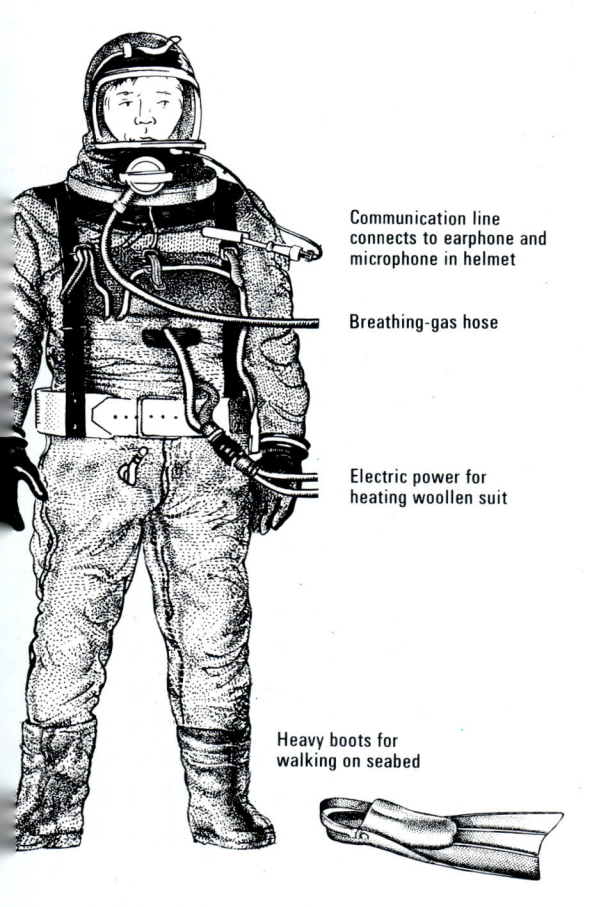

Left: Layers of clothing (one electrically heated) worn by divers working at considerable depths, where cold kills unprotected bodies. Tests by British naval divers have shown that hands need relatively more heat or insulation than any other part of the body. Feet and upper arms come next; then head, torso, and lower arms. Thighs and legs tolerate cold better than the rest of the body.

Below: A cutaway view of 'JIM' – a pressure-proof magnesium alloy diving suit. Inside this metal monster a diver 400 metres (1300 feet) down can breathe air at atmospheric pressure. Thus even after a long dive he can surface safely yet fast, without wasting time decompressing. Articulated limbs equipped with claws enable the diver to perform many tasks.

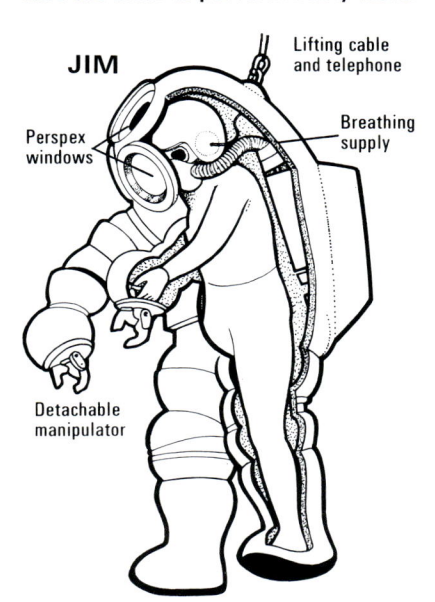

Problems of Pressure

Under pressure, the oxygen and nitrogen in air dissolve in the diver's blood. Too much oxygen may mean convulsions and death; too much nitrogen makes him light-headed and irresponsible, dangerously impairing his judgement.

As the diver ascends, water pressure diminishes. If he rises too fast, the dissolved gases in his blood expand and form bubbles. Bubbles in his joints produce the pains known as 'the bends'. Bubbles in the blood vessels of his lungs cause chest pains and coughing, known as 'the chokes'. Bubbles in the nervous system cause a form of paralysis known to divers as 'the staggers'.

France's Paul Bert (1833–86) and Britain's John Scott Haldane (1860–1936) helped to solve the mystery of the strange diseases that affected divers. Bert showed how gas under pressure can turn a man into a fizzing human champagne bottle. Haldane worked out tables showing how fast a diver can safely rise according to the length and depth of his dive. Even brief dives to great depths call for a long, slow return to normal atmospheric pressure. As the graph shows, 1 hour's work at 120 metres (385 feet) may mean 27 hours spent rising by stages to the surface or in a DECOMPRESSION chamber.

Deep Diving Today

Haldane's tables made deep diving safer. So did substituting helium for nitrogen in BREATHING MIXTURES. By the mid 1970s experimental divers were plunging to the once unimaginable depth of 450 metres (1476 feet).

There have been more improvements. In 1943 Jacques-Yves Cousteau and Emile Gagnan devised the Aqualung, a SCUBA system. Its frogmen users breathe air from cylinders on their backs and swim unencumbered by airline or heavy boots and helmet. Now, too, articulated metal diving suits like JIM permit men to work deep down at atmospheric pressure and to rise quickly and without the need for decompression.

Underwater Habitats

Left: This old engraving shows the use of a diving bell – an underwater workroom of a type reputedly first used more than 2000 years ago. In this scene, a barrel supplies air by tube to two men in the bell (shown as though transparent). Air from the bell is then breathed in by a diver working on the seabed to recover items from a shipwreck.

Below: An SDC (a modern diving bell) in use. A crane lowers the SDC into the sea from an oil rig. A diver who has undergone compression swims from the SDC, breathing air piped from the SDC. After work, he re-enters it. The SDC is raised, to be locked onto the pressurized DDC. The diver can enter this to rest between dives or for decompression.

In 1870 Jules Verne's *Twenty Thousand Leagues Under the Sea* fancifully described men at home on the seabed. By 1970 men had actually spent weeks in and around undersea dwellings.

Diving Bells

The idea of rooms beneath the waves probably dates from the time of Alexander the Great. More than 2300 years ago, the great Macedonian general supposedly explored the depths in a colimpha – a hollow, man-made shell. In the 16th century AD Greek divers demonstrated a giant bell in which men might breathe and work on the seabed. In 1690 Edmund Halley improved on earlier devices. His diving bell had a piped air supply, enabling men to work longer under water.

Modern diving bells are called SDCs – short for submersible decompression cham-

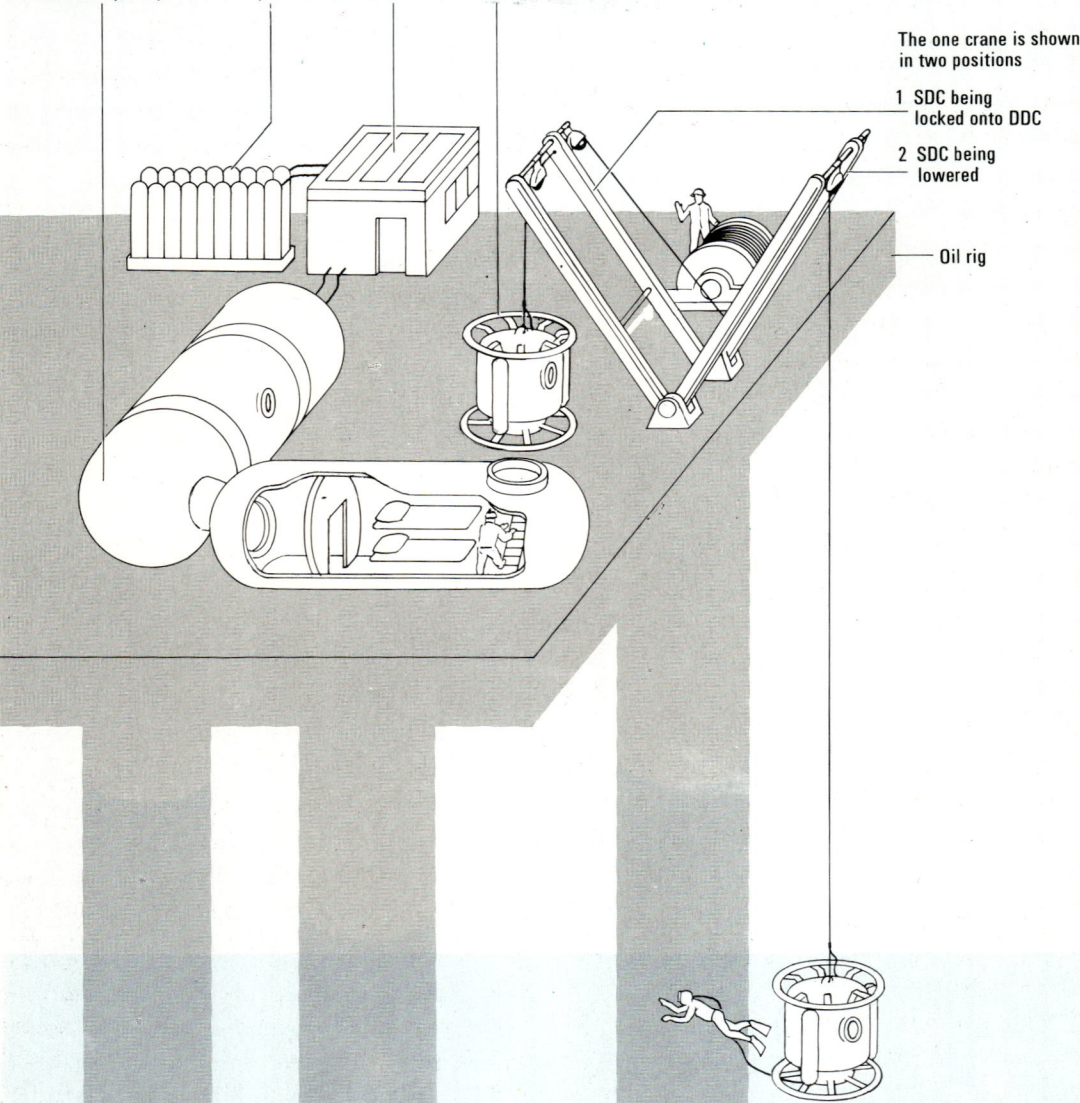

Below: Aquanauts around an underwater home planted in shallow water off the Virgin Islands. *Tektite* is used by scientists for marine and fisheries research. It consists of two steel dwellings connected by a corridor. Each dwelling has two floors. Upstairs are control centre, laboratory, store and observation point; downstairs, crew quarters and 'wet' room. Divers come and go through a hole in the base.

THE SDC-DDC SYSTEM

- Deck Decompression Chamber (DDC)
- Gas cylinders
- Control centre
- Submersible Decompression Chamber (SDC)

The one crane is shown in two positions
1 SDC being locked onto DDC
2 SDC being lowered
Oil rig

bers. Unlike old diving bells, SDCs have breathing mixtures at pressures that can be made to match those of the surrounding water. Attached to a submerged SDC by a breathing tube, divers can work outside for hours at considerable depths. Re-entering the SDC, they are winched up to a ship or platform on the surface, where they crawl into a DDC (deck decompression chamber) at the same pressure as their SDC. Here, they can eat and sleep, then make another dive without decompression and recompression.

This process exploits a technique called SATURATION DIVING. Within a few hours at any depth requiring breathing mixtures body tissues become saturated with the helium gas used in them. The saturated diver can stay indefinitely at that pressure without adding to the final decompression time.

Living Under Water

In the 1960s, saturation diving played a key role in the development of UNDERWATER HABITATS – pressurized bases where divers could eat and sleep for days while working on the sea-floor. The idea seemed attractive. Instead of being confined to the cramped quarters of an SDC–DDC system, the divers could make themselves more comfortable in a larger space under water.

In 1962, France's Jacques-Yves Cousteau proved that such a base was possible. His *Conshelf 1* was a giant steel barrel tethered 10·5 metres (34 feet) down off Marseille. It held two bunks and breathing apparatus. Aquanauts Albert Falco and Claude Wesley lived here for a week, eating food brought down by other divers and working up to five hours daily at depths down to 26 metres (85 feet). They had won a temporary foothold on the continental shelf.

Cousteau's first success inspired him to plan systems featuring a shallow base camp and a succession of deeper ones. In the Red Sea within a year he set up *Conshelf 2*: two underwater homes and hangars for underwater vehicles. In 1965 six so-called oceanauts breathing heliox (a 98 per cent helium, 2 per cent oxygen mixture) worked for more than two weeks from *Conshelf 3* – a two-storey home inside a pressure sphere, at a depth of 128 metres (420 feet) in the Mediterranean. The future looked promising.

Meanwhile, other nations had been busy. In the 1960s, West Germany, Japan, Russia and the United States all tested underwater habitats. By 1969 America's *Sealab 3* was poised to beat the *Conshelf* depth and endurance record. Then the US Navy called a halt.

The High Cost of Low Living

By 1970 the idea of the underwater house began to seem a nine-day wonder.

There were several reasons for this decline in interest. They included the high cost of building and maintaining underwater houses, and the impossibility of supplying them in storms, which sometimes rage for weeks. Then, too, a fixed underwater base ties divers to a small area.

All these snags are avoided by the lockout submersible – a pressure chamber built into a submarine. Nonetheless, underwater habitats may be revived for future underwater mining operations and biological research, particularly in shallow waters.

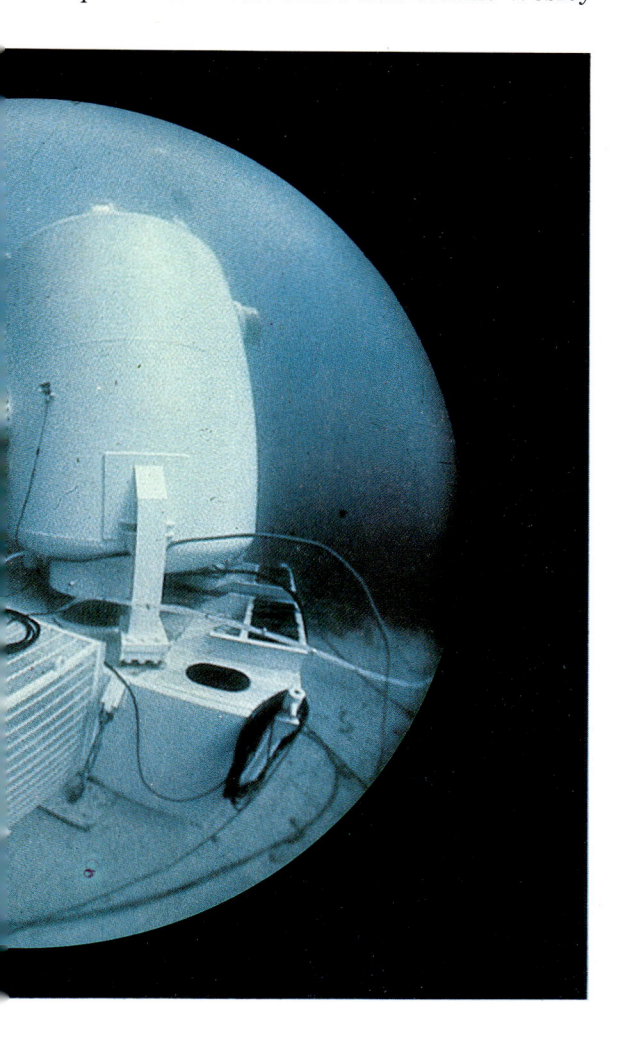

DIVER LOCKOUT SUBMERSIBLE

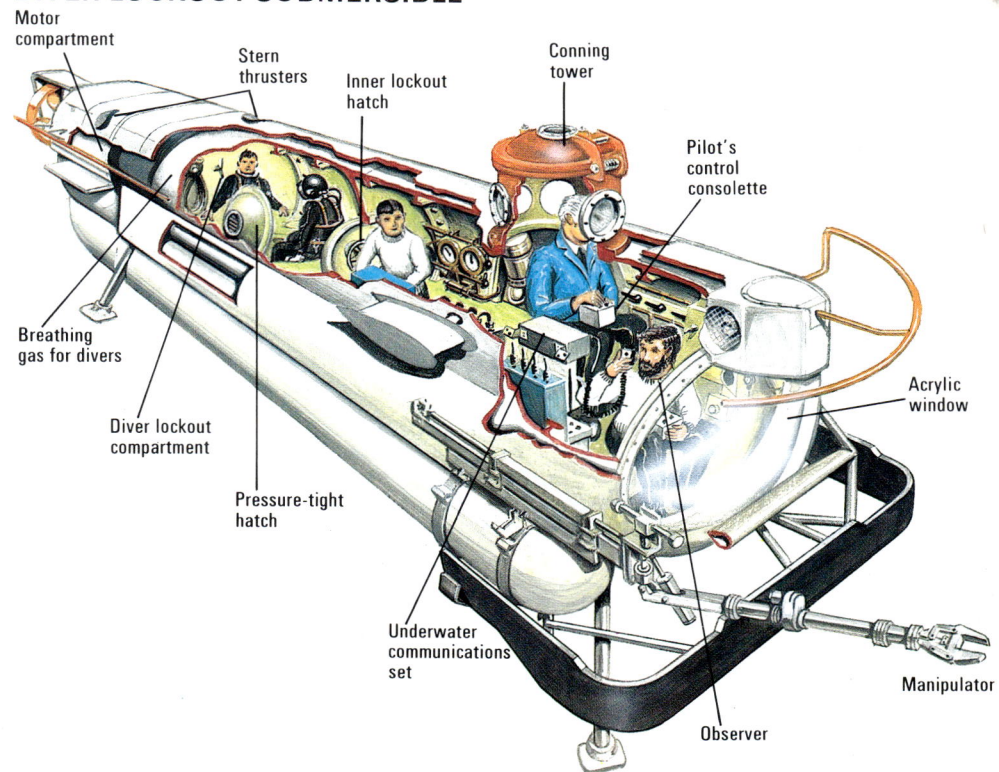

Above: Exploded view of a lockout submersible – a two-part vehicle. The crew travel in a forward cylinder at atmospheric pressure. The divers sit in a separate rear cylinder at the pressure of the depth at which they will emerge and work. Such mobile bases have helped to make the underwater house idea redundant.

Working Under Water

Divers say that any task done on land can be performed under water. Divers and underwater vehicles between them help to survey sites; blast navigation channels; lay pipes and cables; build tunnels and bridges; inspect and clean underwater structures and protect them from corrosion; repair underwater chains and pipes; and salvage sunken vessels.

Working under water is not easy. Many divers labour in cloudy conditions, encumbered by an airline, cold fingers, and the weightlessness that makes it difficult to use powered tools.

Underwater Tools

There are now devices and techniques to aid work under water. Surveyors plotting sea-floor pipe routes or foundation boundaries may use laser rangefinders emitting green light that penetrates water. In murky water, though, they must rely largely on compass and tape measure. To sample seabed rock, a diver uses a power drill. And to prevent this whirling his almost weightless body around, he tethers himself to the worksite by clamps.

Some tools help divers protect underwater structures from decay and damage. For example, paint pumped down to a brush by tube enables a diver to give underwater pipes a corrosion-proof coat. Powered scrubbing brushes strip encrustations off ships' hulls.

Damage happens in spite of such precautions. Iron rusts in seawater. Waves cause metal joints to vibrate and crack. Fishing nets and ships' propellers foul submarine cables. Divers must make regular inspections. Waterproof television cameras and ultrasonic sensors help to reveal corrosion and hidden cracks.

A large armoury of tools and substances aids underwater repair and construction. Divers use hydraulically powered rotating saws to cut free cables which are fouling ships' propellers. They can burn their way through metal with a blow torch whose air jet protects the torch's burning gas from the surrounding water.

To mend a damaged concrete-surfaced metal pipe, divers clean the 'wound' then dress it with an epoxy resin patch, smoothed down by a kind of rolling pin.

Underwater tools have produced remarkable results. Between 1917 and 1924 British divers salvaged 3186 of 3211 gold bars lost when the *Laurentic* sank off Ireland. Today, commercial divers routinely mend pipelines in far deeper waters.

Machines on the Seabed

Machines as well as tools are fast transforming work beneath the sea. Self-propelled submersibles play valuable roles (see pages 54–55), but often low power limits their capacity. Much inspection work is now being done by remote-controlled machines. New machines such as the tracked, bottom-crawling bulldozer and trench digger are now being developed to be the workhorses of tomorrow, along with vessels capable of manipulating small pieces of equipment.

Above: A diver operating a mobile drilling rig under water.

Right: Divers attending a pipeline in shallow Caribbean waters warm enough for them to leave their legs unprotected from the cold. Pipeline work includes repair to oil lines fouled and fractured by fishing gear and anchors. Cold and murk can make such operations difficult.

Below: Sparks fly under water as a diver burns through a cable. Divers can use a blow torch for such work in relatively shallow water. But at great depth, water pressure acting against the gas jet forces welders to use electric current instead.

Seabug is one of the new breed of remotely controlled subsea vehicles. It can cut, bore, lift, dig a trench, survey, measure, inspect pipelines, sample soil, etc.

Submarines and Submersibles

Cornelis Drebbel's leather-skinned, wooden-hulled SUBMERSIBLE was rowed beneath the Thames in 1620. But it was not until the late 19th century that SUBMARINES reached considerable depths and could travel a useful distance submerged with enough air for the crew to breathe.

Military Submarines

Progress came at first with military submarines like John P. Holland's *Holland*. Accepted by the US Navy in 1900, this prototype used petrol engines at the surface, and battery-powered electric motors under water, which operate without the use of air. But diesel-electric submarines must recharge their batteries by surfacing to run their engines, or by drawing air in for the engines through a breathing tube or snorkel.

The next step forward in propulsion came in 1955 with the US Navy's nuclear-powered *Nautilus*. Heat from its reactor yields steam for driving turbines. The engines need no air, and the crew breathe air chemically recycled or drawn in by snorkel. By the 1970s strong-hulled submarines like the USS *Trident* could stay 1000 metres (3280 feet) down for months. If necessary they can circumnavigate the world submerged at speeds of 30 KNOTS or so. Also, military submarines have a greater capacity for destruction than ever before. In World War II, submarines could sink surface vessels by torpedo. Today, they can fire nuclear missiles at cities 7800 kilometres (4875 miles) away.

Below: An American nuclear submarine designed to fire nuclear Polaris missiles while submerged. A single American submarine can now fire 224 nuclear warheads in five minutes at targets thousands of kilometres away.

Submersibles

Military submarines are only part of the picture. Since the late 1950s, a whole new family of mini-submarines has been designed for underwater work and study.

First of these submersibles was Jacques-Yves Cousteau's *Soucoupe*, or diving saucer. Strong yet light, its hull was planned for probing the seabed and its life down to 300 metres (1000 feet), a considerable depth in 1959.

In the mid 1960s, the United States mobilized several deep-sea research and recovery submersibles to hunt for a hydrogen bomb lost in the sea off south-east Spain. One was the aluminium-hulled *Aluminaut*, built to take three men and a tonne of gear down 4600 metres (15,000 feet), and to pick objects off the seabed with grappling arms in its nose. The smaller *Alvin* had a crew of two and was well endowed with observation ports and navigation aids, including lights and sonar. In fact the *Alvin* found the bomb on a slope 950 metres (3100 feet) down, but failed to grapple it. Retrieval came with a third craft, the US Navy's cable-controlled unmanned vehicle – known for short as CURV.

By the late 1970s, civilian submersibles were routinely working under water on cables or with gas and oil development on the continental shelf off Europe, North America and Asia. Many have more power and dive deeper and for longer periods than precursors like the diving saucer. Some remote-controlled submersibles are also used.

Above: Jacques-Yves Cousteau peers from an observation porthole in the so-called *Sea Flea*. Lights aid underwater exploration by this small submersible, one of a whole family of manned underwater probes.

Bottom: Four cutaway views show how a submarine dives and rises. Air in ballast tanks gives enough buoyancy to keep the vessel floating. When some air is released, water flows in, the vessel loses buoyancy and sinks. When the diving control officer blows water from the ballast tanks, the vessel rises. Speed and angle of ascent or dive can be controlled by elevating or depressing short horizontal fins called diving planes, and by varying the rate at which tanks fill and empty.

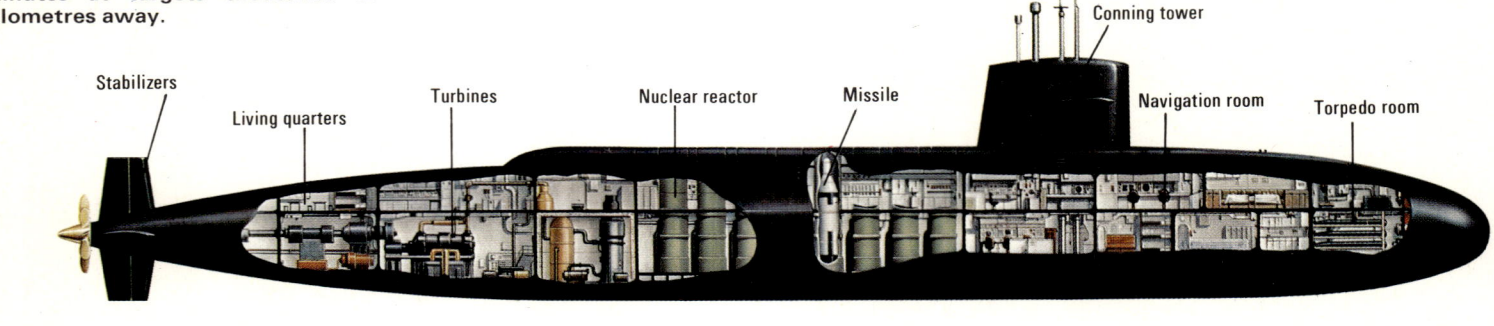

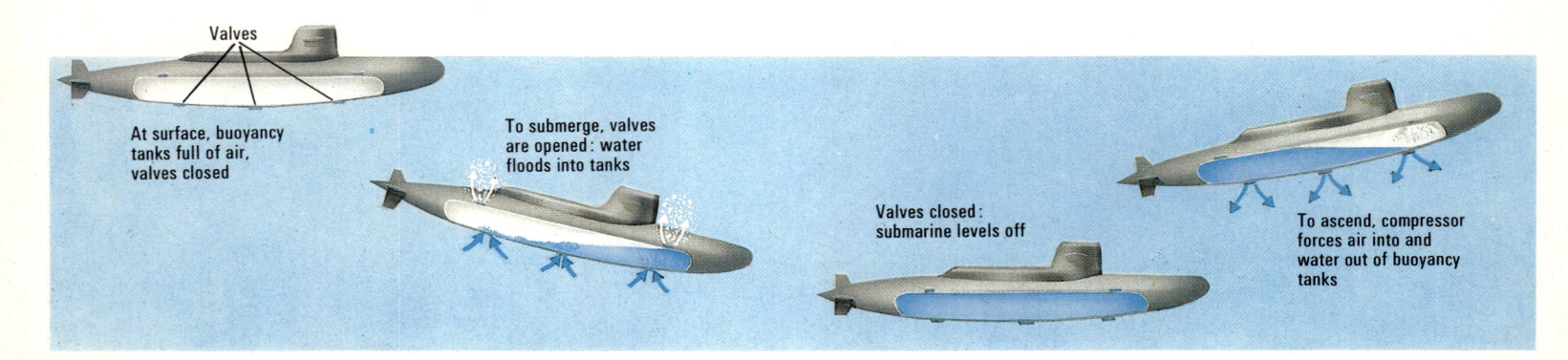

Stabilizers · Living quarters · Turbines · Nuclear reactor · Missile · Conning tower · Navigation room · Torpedo room

At surface, buoyancy tanks full of air, valves closed

To submerge, valves are opened: water floods into tanks

Valves closed: submarine levels off

To ascend, compressor forces air into and water out of buoyancy tanks

Left: Topside view of *Pisces III*, a Canadian-built two-man submersible. An outer sheath of fibreglass protects pressure spheres designed to withstand water pressure 900 metres (3000 feet) down. Crewmen peering from ports in the big forward sphere can inspect subsea pipelines, and gather rock samples by using a mechanical arm.

Below: *Consub*, a submersible workhorse remotely controlled from the surface by a pilot and observer, and electrically powered via an umbilical cable. Cameras, sonar gear, and power tools including artificial hands enable *Consub* to survey, sample and inspect at depths down to 600 metres (2000 feet).

Below: An artificial bubble gives two men an uninterrupted view of the underwater world. Christened *Nemo*, this US Navy observation sphere pioneered the use of acrylic pressure hulls for submersibles when it appeared in 1970. Two years later, *Deep View* incorporated a glass hemisphere in its pressure hull.

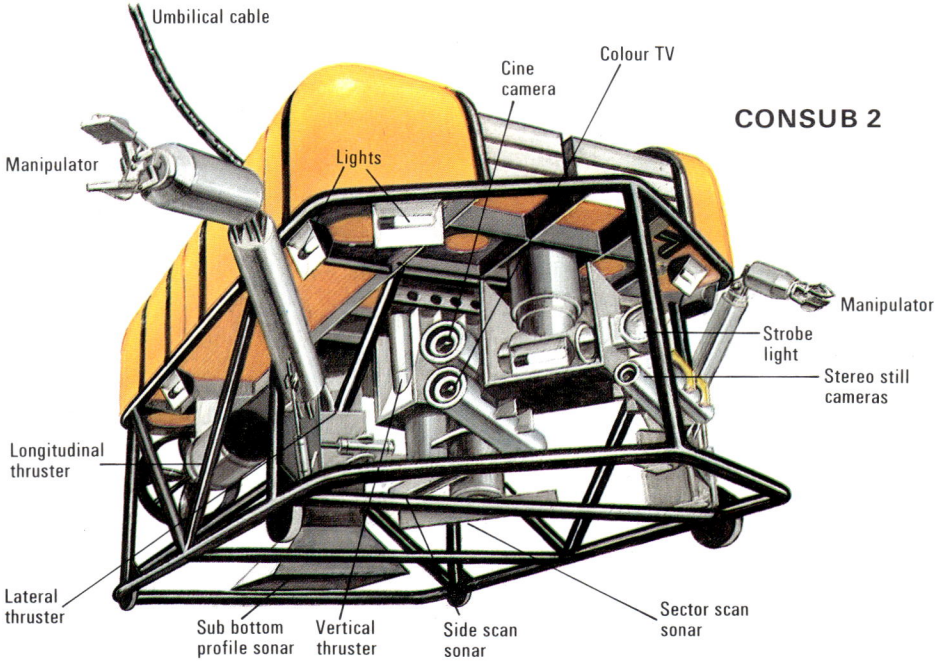

CONSUB 2

- Umbilical cable
- Cine camera
- Colour TV
- Manipulator
- Lights
- Manipulator
- Strobe light
- Stereo still cameras
- Longitudinal thruster
- Lateral thruster
- Sub bottom profile sonar
- Vertical thruster
- Side scan sonar
- Sector scan sonar

Right: RCV is an unmanned submersible equipped with television cameras and used mainly for inspecting underwater installations. It is very small (66 cm by 51 cm) and is relatively cheap and easy to operate.

55

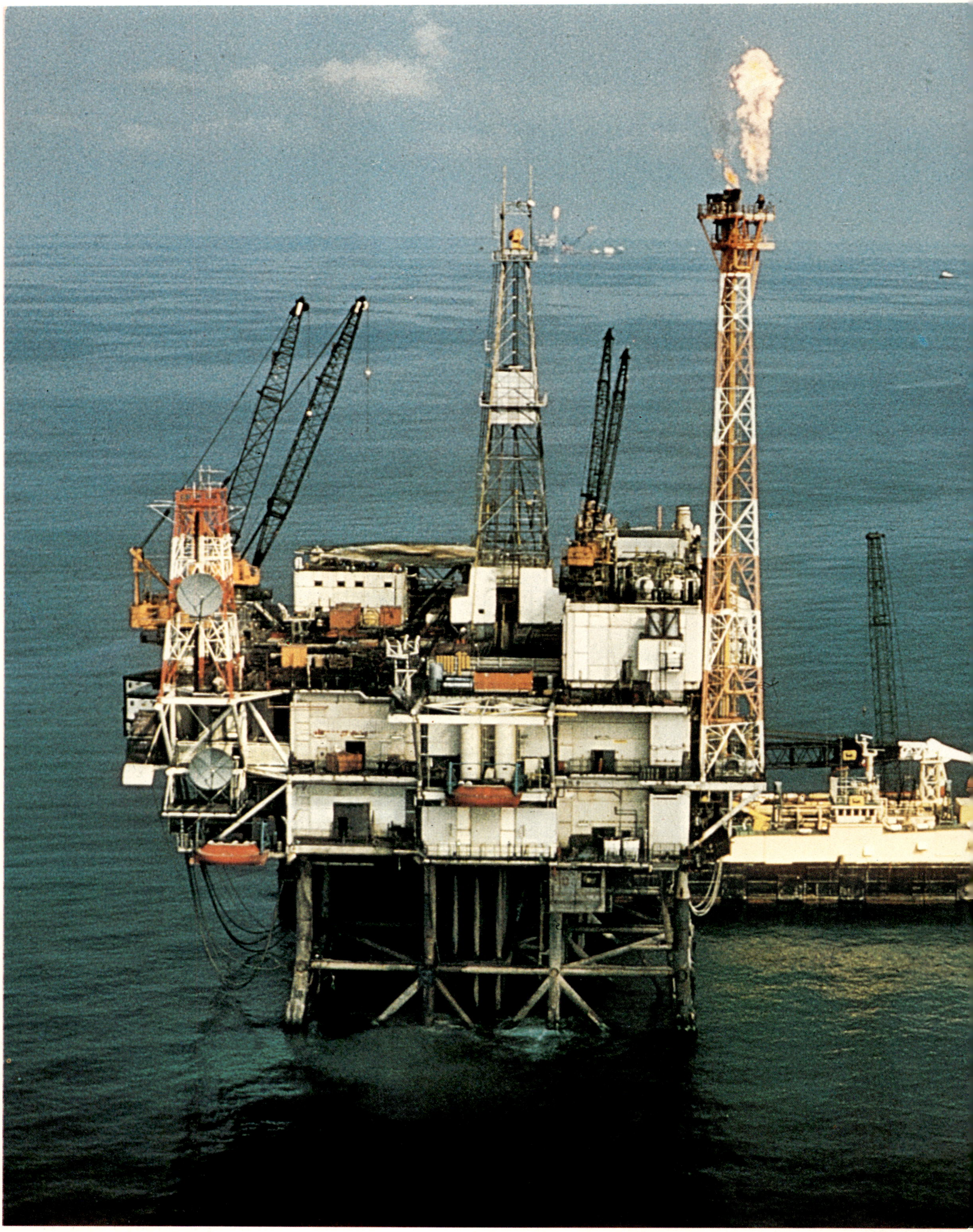

Chapter Five

WEALTH FROM THE OCEANS

Seawater and the seabed contain many potentially useful chemicals. Most of these substances are usually mined and quarried from the land, but industry is fast using up these irreplaceable supplies. Some nations are now beginning to look to the oceans to provide the metals, salts and fossil fuels without which industries and cities would run down.

Separating dissolved substances from seawater can be expensive. Finding and extracting oil deep below the sea is difficult and sometimes dangerous. Solving these problems has meant devising whole new technologies. Yet this is proving well worth while. For instance, tapping seabed resources has postponed the day when oil supplies run out.

But some seabed treasures have historic, not economic, value. Sunken ships and cities now being found by underwater archaeologists throw fresh light upon the daily life of past civilizations.

A drilling rig in the North Sea. Drilling and production rigs float or stand above the continental shelves off most continents. By the mid 1970s undersea oil fields provided one-fifth of the world's petroleum output. Soon they will be providing one-third of all supplies.

Mining the Waters

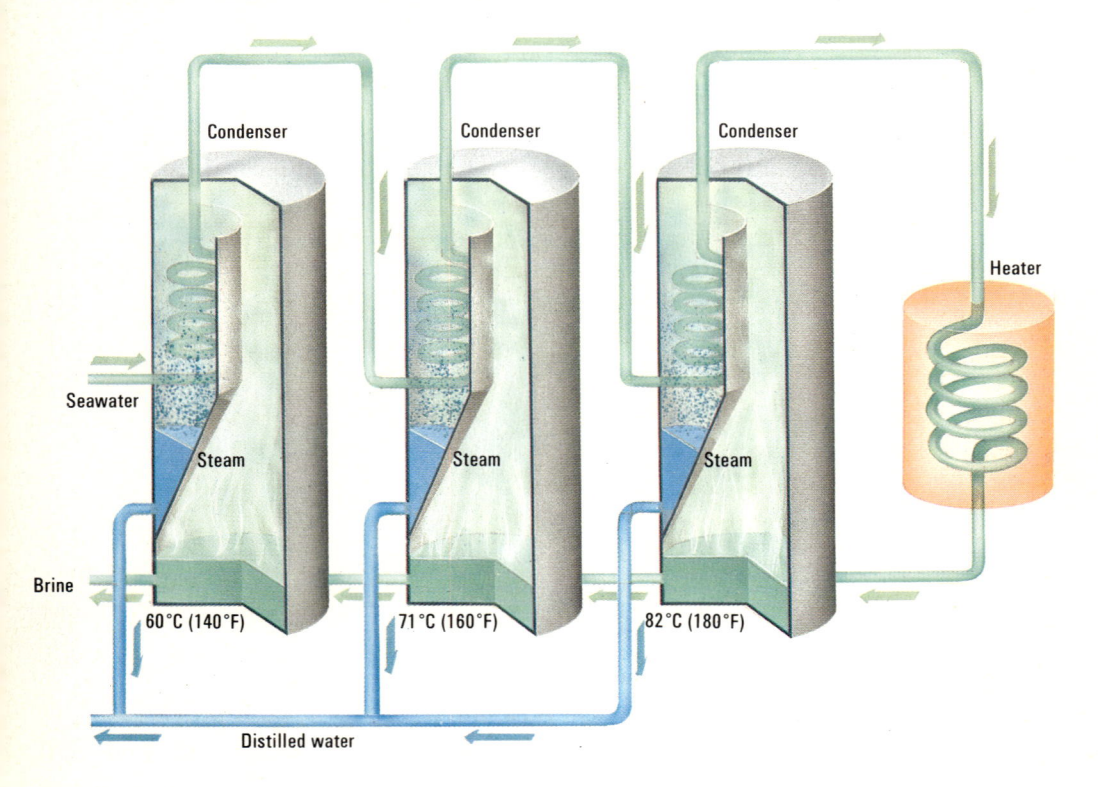

In some ways, seawater is a vast mine. It holds perhaps all the chemical elements found on Earth, some in massive quantities. The only problem is how to concentrate them and separate them from each other.

Water

Of course, water itself is the most abundant substance in the ocean. But the DESALINATION of large quantities of it uses a lot of energy. Even so, this proves worth while in largely hot, arid, but oil-rich, countries – for instance Mexico, Kuwait and Abu Dhabi. Multistage flash DISTILLATION is the usual process. FREEZING salt water and collecting the ice is one of the ways of desalting brine which is still being researched. But this method is even costlier.

Salt

Next to water, salt is the most abundant substance in seawater. Indeed, there is enough to coat the continents 150 metres (480 feet) deep. Separating salt from brine by evaporation is simple, and people have been doing it for thousands of years. The brine is pumped into shallow pools called pans. Sun and wind evaporate the water, until calcium carbonate and calcium sulphate settle in the form of crystals. Workers then move the brine to other pools where sodium chloride ('salt') crystals accumulate. Next, used brine is removed and a fresh, concentrated supply is pumped in. This happens repeatedly until the deposited salt crust is thick enough to gather.

Where wind and sun are insufficient to produce salt quickly, it can be separated from seawater more expensively by boiling or freezing. Of course the cheapest way of getting sea salt is to mine the thick deposits left behind on land when prehistoric seas evaporated. Most of the salt used in industry and cooking in the West comes from such deposits.

Magnesium and Bromine

After salt, magnesium and bromine are the two most valuable substances extracted from the sea. Magnesium is a metal used as an ingredient in strong, light alloys, and to protect pipes and ships' hulls from corrosion. There are 4 million tonnes of magnesium in every cubic kilometre of ocean (about 6 million tons per cubic mile). Oceans supply most of the world's magnesium today. Manufacturers collect it in huge tanks of seawater mixed with lime. The magnesium precipitates out of the water and settles as magnesium hydroxide. This is filtered off, and treated with hydrochloric acid. The result is magnesium chloride, which is fed into an electrolytic cell. There, electric current splits it into chlorine and magnesium.

Above: Salt pans in north-western France. Trapped in shallow, man-made pools, seawater evaporates, leaving salt behind. Workers then gather it in heaps for distribution.

Below: Desalinating seawater by multistage flash distillation. Heated seawater 'flashes' (boils instantly) as it passes through low-pressure chambers. The resulting steam pre-heats cold seawater entering the system. As the steam yields heat it turns to water that runs off to be collected.

Substances extracted from seawater	per cent of world production
Bromine	70
Magnesium metal	61
'Manufactured' water	59
Salt	29
Magnesium compounds	6

MULTISTAGE FLASH DESALINATION

Other processes make seawater the source of about 70 per cent of the world's bromine – a chemical employed in dyeing, medicine, metallurgy, motor fuel and photography.

Bio-accumulation

Gold, silver, tin, titanium and uranium all occur in seawater, but too scantily to make extraction feasible today. However, we know that some marine plants and animals can concentrate rare substances. Such BIO-ACCUMULATION produces pearls in oysters and iodine in seaweeds. We may yet find and farm sea organisms capable of concentrating other useful substances.

Right: An oyster split apart to reveal pearls. Oysters build pearls by coating small foreign bodies such as particles of sand with layers of nacre which they make from substances in seawater. Oysters also accumulate zinc in their body tissues. Other sea creatures concentrate copper, niobium, vanadium – elements scarce in seawater. Such organisms may one day help to satisfy our needs for some rare metals.

Below: A magnesium-producing plant owned by the Kaiser Aluminum and Chemical Corporation in Oakland, California. It produces 150,000 tonnes of magnesia and 165,000 tonnes of magnesium hydroxide a year.

RESOURCES ON THE SEABED

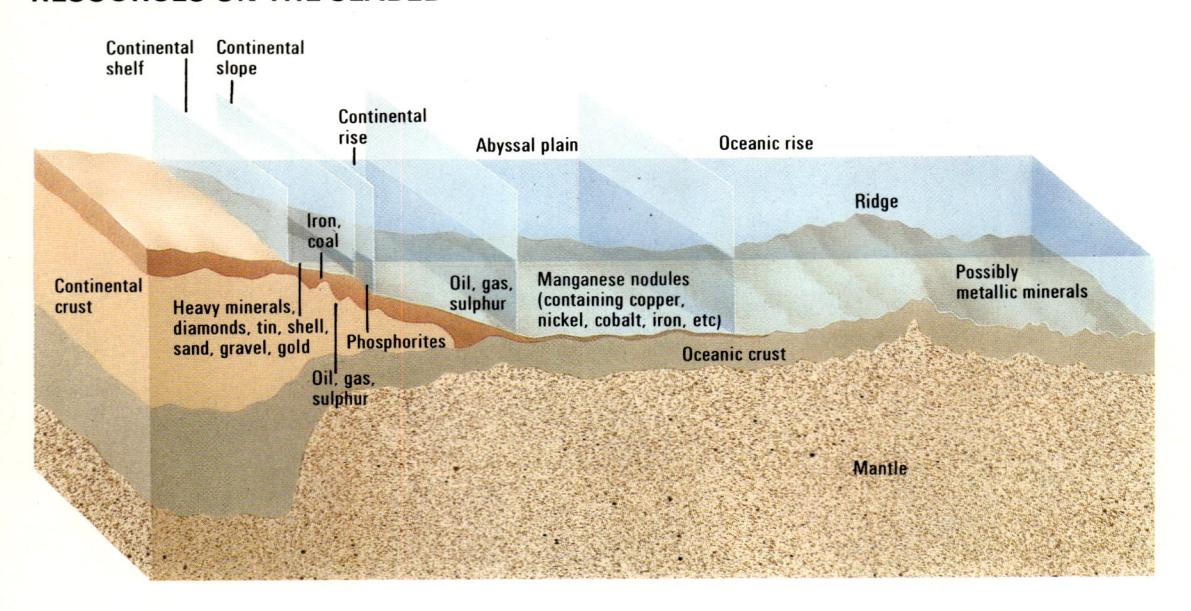

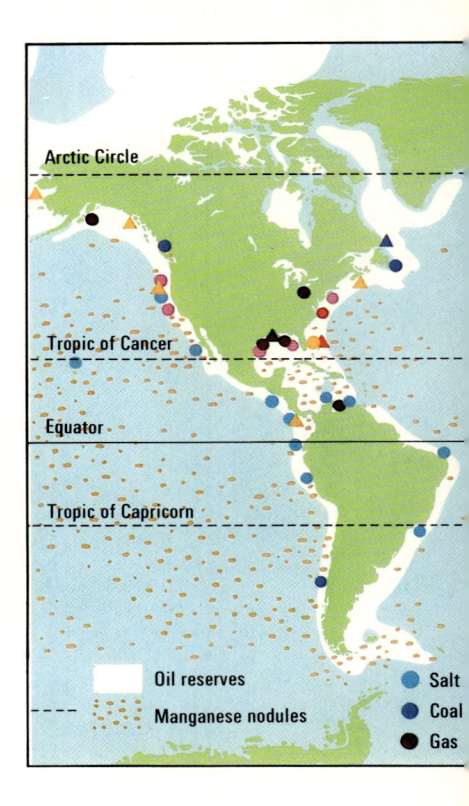

Mining the Seabed

Above: Cutaway view of the ocean floor, showing where different types of mineral resource occur.
Right: A map showing the global distribution of seabed oil, gas and other mineral resources.

CONTENT OF A MANGANESE NODULE

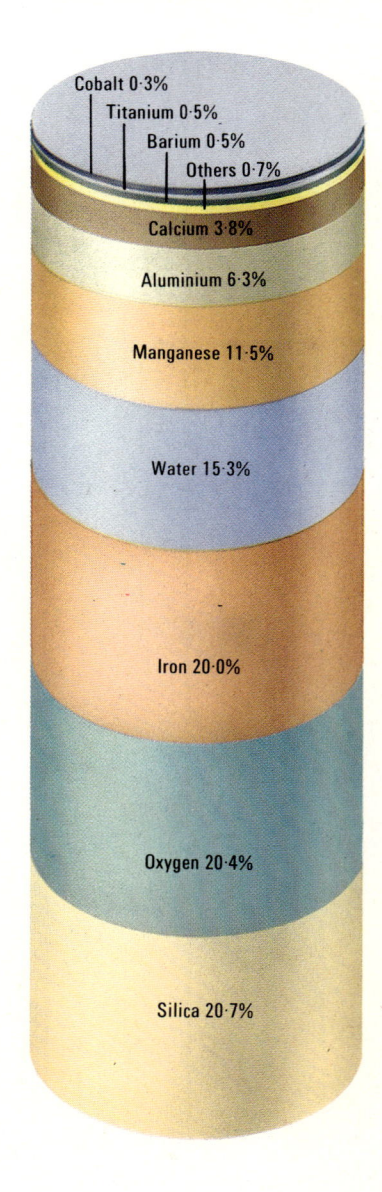

- Cobalt 0·3%
- Titanium 0·5%
- Barium 0·5%
- Others 0·7%
- Calcium 3·8%
- Aluminium 6·3%
- Manganese 11·5%
- Water 15·3%
- Iron 20·0%
- Oxygen 20·4%
- Silica 20·7%

The continental shelf and ocean floor are rich in certain minerals. Some are used already. Some await the right technology. Mining others may never pay.

Many offshore mineral deposits occur in sediments lying on the seabed surface. They include ores weathered from rocks inland, washed into the sea by rivers, and sorted by waves, tides and currents. In this way sands and GRAVELS rich in metals collect off certain coasts.

Offshore Dredging and Mining
Of all the solids waiting to be scooped up from the seabed, sands and gravels are so far most profitable. Off shallow European and American coasts, big diesel-powered pumps in dredging vessels can suck 1800 tonnes of aggregate an hour from the seabed up to 45 metres (150 feet) below. The pumps fill barges capable of holding 4000 tonnes of sand and gravel.

What has been called the world's largest single mining operation involves dredging sands in the Bahamas. These sands are rich in aragonite – a form of calcium carbonate used in glass and cement. Aragonite also enriches animals' diets which are deficient in calcium. The sands are sucked up onto a man-made island and shipped by bulk carrier to Florida and other markets.

In some parts of the world, people dredge for minerals more highly priced than these building materials. Burma, Indonesia, Malaysia and Thailand have deposits of cassiterite, a substance rich in tin. Dredgers scoop it from channels up to 8 kilometres (5 miles) offshore. Farther west, dredgers get barium sulphate from Sri Lankan waters up to 1200 metres (4000 feet) deep.

Diamonds, gold and phosphorite are other minerals found in offshore sediments. But prohibitive expense in 1972 halted diamond-dredging off Namibia.

Underwater prospecting problems, high extraction costs and bad weather have made offshore mining far harder than the optimists expected in the 1960s. Because of this, few known deposits are actually worked, apart from sands and gravels.

Mining hard rock under water is another matter. Shafts begun on land and driven up to 6 kilometres (4 miles) beneath the sea off north-east England yield rich coal supplies. Others could yield potash.

Dredging the Abyss
If dredging inshore waters has largely proved impracticable, dredging the deep ocean floor might seem impossible. In fact it could well prove worth while. The main attraction is millions of rounded, metallic lumps scattered on the seabed – especially in the Pacific Ocean. These lumps are called MANGANESE NODULES, but they also contain cobalt, copper and nickel and other substances whose land deposits are rapidly dwindling.

Geologists believe that the nodules grow as substances precipitated from the water collect around objects, such as fish teeth, lying on the seabed. The nodules range in size from pebbles to rocks that weigh a tonne. They grow slowly – probably by only 0·1 millimetres in a thousand years. But this is actually faster in total than the rate at which we use some of the metals that they contain.

Dredging useful quantities from great depths poses major problems. So does separating the ingredients, whose useful contents vary in amount and are lower than people once believed. By the late 1970s technologists had made much progress, but mass dredging had not started, and it may be years before it can be made to pay.

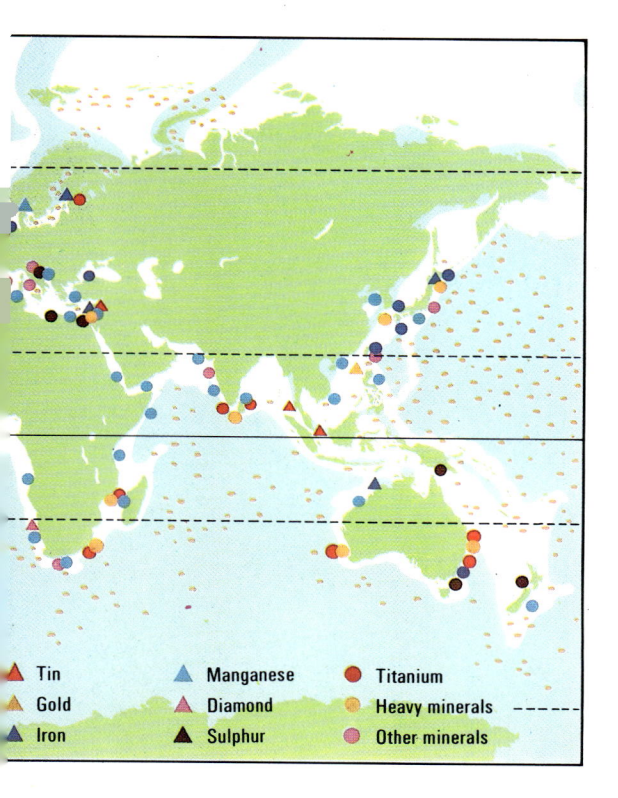

A dredger at work. Dredgers like this are used for mining sand, gravel, and mineral-rich sands in shallow coastal waters as well as for excavating navigational channels.

Offshore mining in 1970	thousands of tonnes
Sand and gravel	55,000
Calcium carbonate	18,800
Sulphur	1000
Barytes	122
Iron sands	36
Tin	12·5

▲ Tin
● Gold
▲ Iron
▲ Manganese
● Diamond
▲ Sulphur
● Titanium
● Heavy minerals
● Other minerals

Left: Proportions of various elements found in a manganese nodule, based on samples from the Atlantic ocean bed. (The contents of individual nodules vary greatly.) Raising nodules in bulk and extracting their useful ingredients offers one of the major challenges in ocean technology.

Three ways in which manganese nodules may be collected from the deep ocean floor. The ship on the left 'vacuums' them off the bottom. The two ships in the middle pull the continuously moving series of dredge buckets that scoop up the nodules. The system on the right relies on remote controlled collectors that shuttle between the ocean floor and the surface platform.

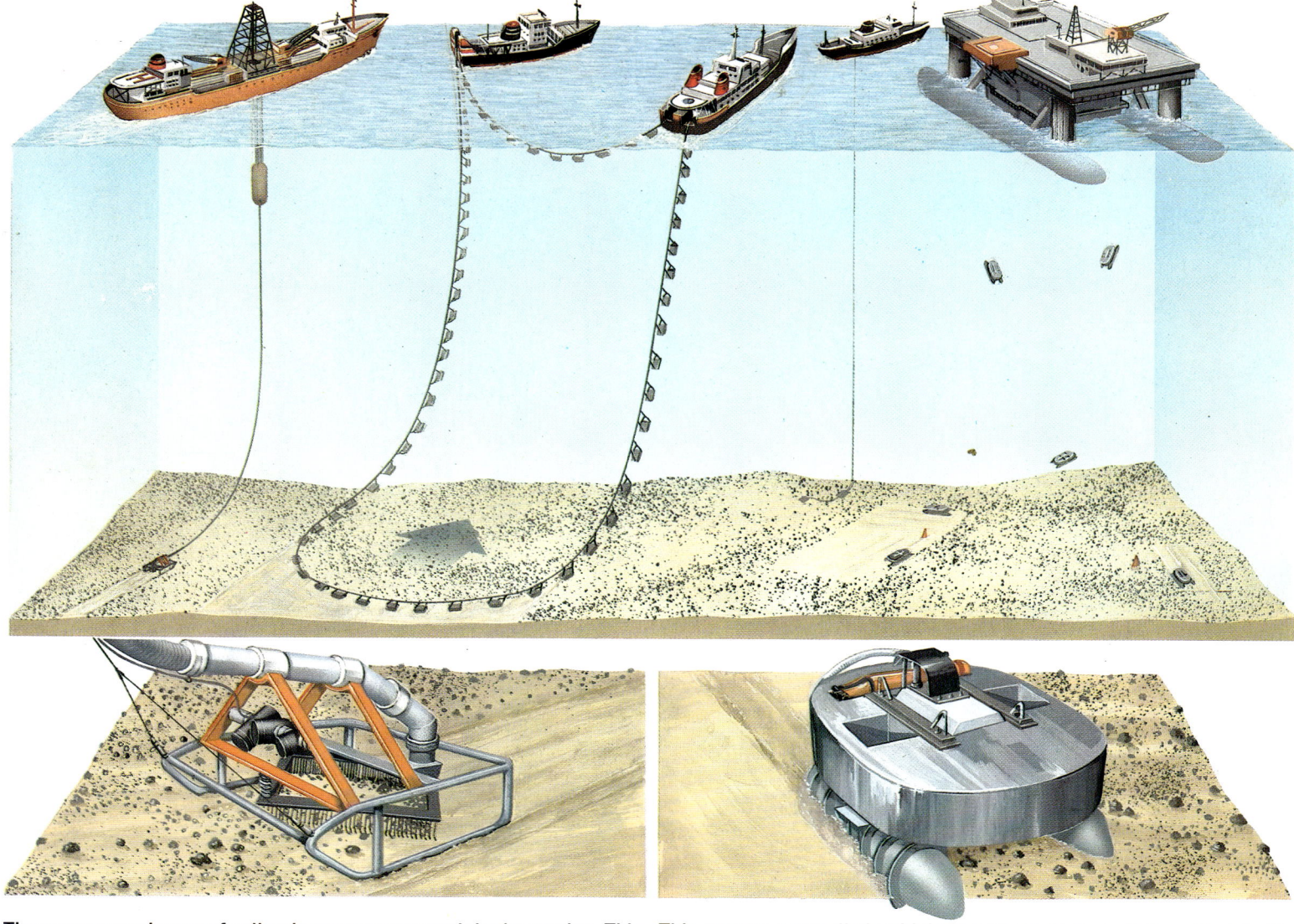

There are several ways of collecting manganese nodules by suction. This sledge has a rake to scrape the nodules off the ground before being sucked to the top.

This remote controlled vehicle moves on screws. It is guided to the mining area by a sonar homing beacon. A factory platform processes the nodules brought to the surface by these shuttles.

Oil and Gas

Below right: An artist's impression based on a subsea oil production system envisaged by the American company, Lockheed. The oil is drilled in several wellhead cellars and piped via the manifold centre to the oil production platform where it is processed ready for transport to the shore. It may be stored in a huge underwater tank and taken to the shore by oil tanker, or it may be transported direct by pipeline.

The continental shelves are only one-sixth the size of dry land, but hold one-quarter as much rock likely to contain oil or natural gas. Technologists have been mining seabed oil since the 1890s. Originally this was carried out mainly from structures similar to those used on land. Modern technology now exploits oil and gas resources far offshore.

Finding Oil and Gas

Oil and gas do not occur just anywhere. They are found where silt or sand settled on the seabed, burying plant and animal remains. The sediments hardened into rock. Heat, pressure and bacteria acting on the dead, crushed organisms converted them to HYDROCARBONS. It is where these lie trapped in porous rocks that sub-seabed reservoirs of oil and gas occur.

Finding oil fields and gas fields under water is tricky. Local variations in gravity or the Earth's magnetic field help geologists to pinpoint likely areas. Echoes from underwater explosions give some clues to underlying rocks. But only a test drilling can prove that oil or gas is actually there.

Test Drilling

Early offshore exploration involved drilling from piers. Then came mobile rigs. JACK-UP PLATFORMS combine a hull and legs. Tugs tow

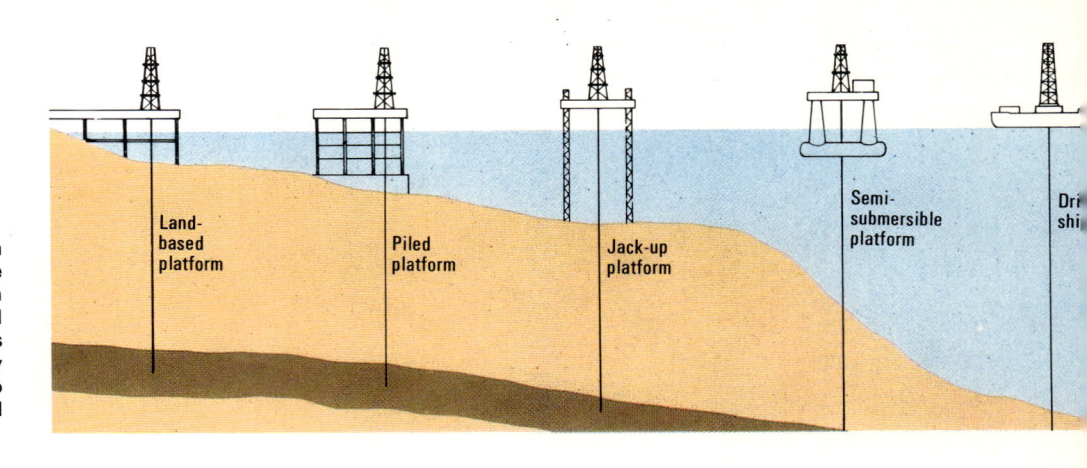

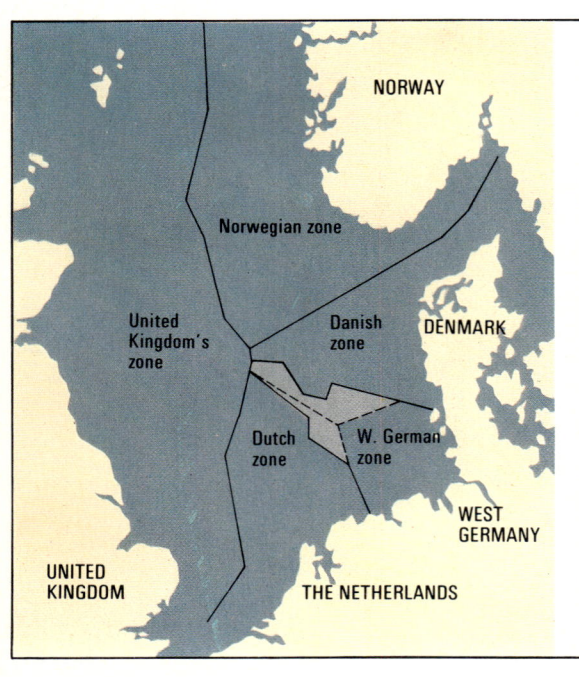

Left: How nations bordering the North Sea agreed to share out its mineral resources. The United Kingdom and Norway gained the lions' shares. Denmark, West Germany and the Netherlands acquired much smaller sectors. The original division, based on the length of each country's coastline, and shown by the dotted lines, gave West Germany an even smaller area. This was adjusted by international agreement. When the North Sea 'cake' was cut in the 1960s, no one knew how much wealth in gas or oil might lie below its floor. One big hint had been a large Dutch mainland gas discovery in 1959. Ten years passed before the first big oil strike. Exploration showed that British and Norwegian sectors had the richest oil reserves, lying close to both sectors' common boundary. By the late 1970s continued exploration suggested that the North Sea's oil reserves were the world's tenth largest.

Left: The kind of drilling rig used in the search for oil depends on the depth of water it has to operate in.

Offshore oil production in 1975	millions of tonnes
Latin America	99·8
Middle East	54·2
North America	45·3
Africa	41·5
S.E. Asia	26·3
West Europe	17·8
Others	32·0

An oil production platform being towed by tugs to its site in the North Sea where it will be lowered into position.

the hull into position, then the legs are lowered to the seabed. Water more than 40 metres (130 feet) deep calls for semi-submersibles or drillships. A SEMI-SUBMERSIBLE PLATFORM has legs buoyed up by pontoons. On site, ballast tanks are flooded to submerge the pontoons with seawater. Anchors hold the rig in place. Big North Sea semi-submersibles stay stable in 100-knot gales, and can drill down through 9000 metres (30,000 feet) of rock while tethered 300 metres (1000 feet) above the seabed. Drillships kept in place by computer-controlled propellers can operate in water 400 metres (1300 feet) deep. The drill is lowered through a hole in the middle of the ship.

Test drilling involves boring a hole in sections. These are lined with steel and concrete. Chemical 'mud' is pumped down to keep the drill bit cool and remove rock chips. It also helps to prevent a 'blow-out' caused by gas or oil gushing up. Valves and rams can also be used to close off the well.

Most holes prove dry. Fruitful holes are plugged and left for later exploitation by production platforms.

Production Platforms
These costly structures have drilling derricks, living quarters, and generators. Giants built for the deep, wild waters of the North Sea include 30,000-tonne steel platforms on steel piles. Mammoth concrete structures weighed down by ballast stand in water deeper than 200 metres (650 feet).

Helicopters and workboats supply men and materials, and one rig can drill 60 slanting holes to tap an area 3 kilometres (2 miles) across. The oil or gas recovered is piped to a tanker, or ashore. New forms of unmanned subsea systems are being developed, as shown in the illustration.

By the mid 1970s, offshore oil accounted for one-fifth of the world's supply. It is estimated that by the mid 1980s it will provide a third.

Sunken Treasure

Hundreds of sunken ships lie where they were wrecked centuries ago off European coasts and in the Caribbean Sea. Countless Stone Age settlements were drowned when ice sheets thawed and ocean levels rose. Earth movements have plunged many ancient Mediterranean cities under the waves.

Waves or earthquakes have badly damaged many sunken ships and cities. But provided they are not plundered, they can still show us how people used to live.

Fishermen and sponge divers discover some sites by accident. Others are found through systematic hunts by teams of frogmen. Side-scanning sonar and devices known as magnetometers are modern aids to underwater searches.

A site is often difficult to recognize. Encrusting underwater growths and seaweeds may conceal the object's shape. Old wooden ships are usually flattened, badly gnawed by sea creatures, and partly sunk in sand or mud, or strewn among rocks.

Underwater Archaeology

To make sense of what has been discovered, the underwater archaeologist must clear the debris, and plot and label objects where they lie. A grid of tapes or metal bars laid out above the site aids mapping and photography. Air-filled bags help divers raise heavy finds like cannon or stone coffins. Lighter objects go in baskets drawn up by line.

Above, people clean and preserve the finds. Fragile wood is kept in water, then treated with a waxy preservative. Electrolysis can help repair the rusted surfaces of metal objects.

A single underwater excavation may last several seasons. Four summers and five thousand dives went into the first full excavation of an ancient ship, off southern Turkey in the early 1960s. Restoring and studying the finds can take many more years.

What We Have Learned

Underwater archaeology has taught us a great deal, for example about how those great Bronze Age traders the Phoenicians built their merchantmen and what Roman and Byzantine trading vessels looked like. We know, too, how men built the ancient port of Tyre in the Lebanon, and what the busy Roman port of Apollonia in Libya must have looked like in its heyday. In 1692 an earthquake drowned and shattered Port Royal in Jamaica; but divers using airlifts (underwater 'vacuum cleaners') have sucked up thousands of clay, glass, and iron objects – all clues to life three centuries ago.

Left: Lens-eye view of a successful quest for sunken treasure. A diver using a kind of seabed vacuum cleaner clears debris from some of 1000 gold pieces of eight from a ship lost off the Shetland Islands. Dozens of Spanish vessels sank rounding Scotland and Ireland after the Armada was defeated and dispersed in 1588.

Above: Amphoras (wine jars) from an ancient wreck off Bodrum, southern Turkey. Underwater archaeologists put up the metal grid in order to help them plot the positions of such finds.

Right: Two diver archaeologists surveying an underwater site. The waterproof instruments they carry enable them to make notes actually on the spot – a valuable aid to recording the precise location of every object that they find.

Below: To shift rocks from a wreck site this diver loads them into canvas bags, and fills the bags with air from cylinders. The air lifts bags and rocks above the seabed.

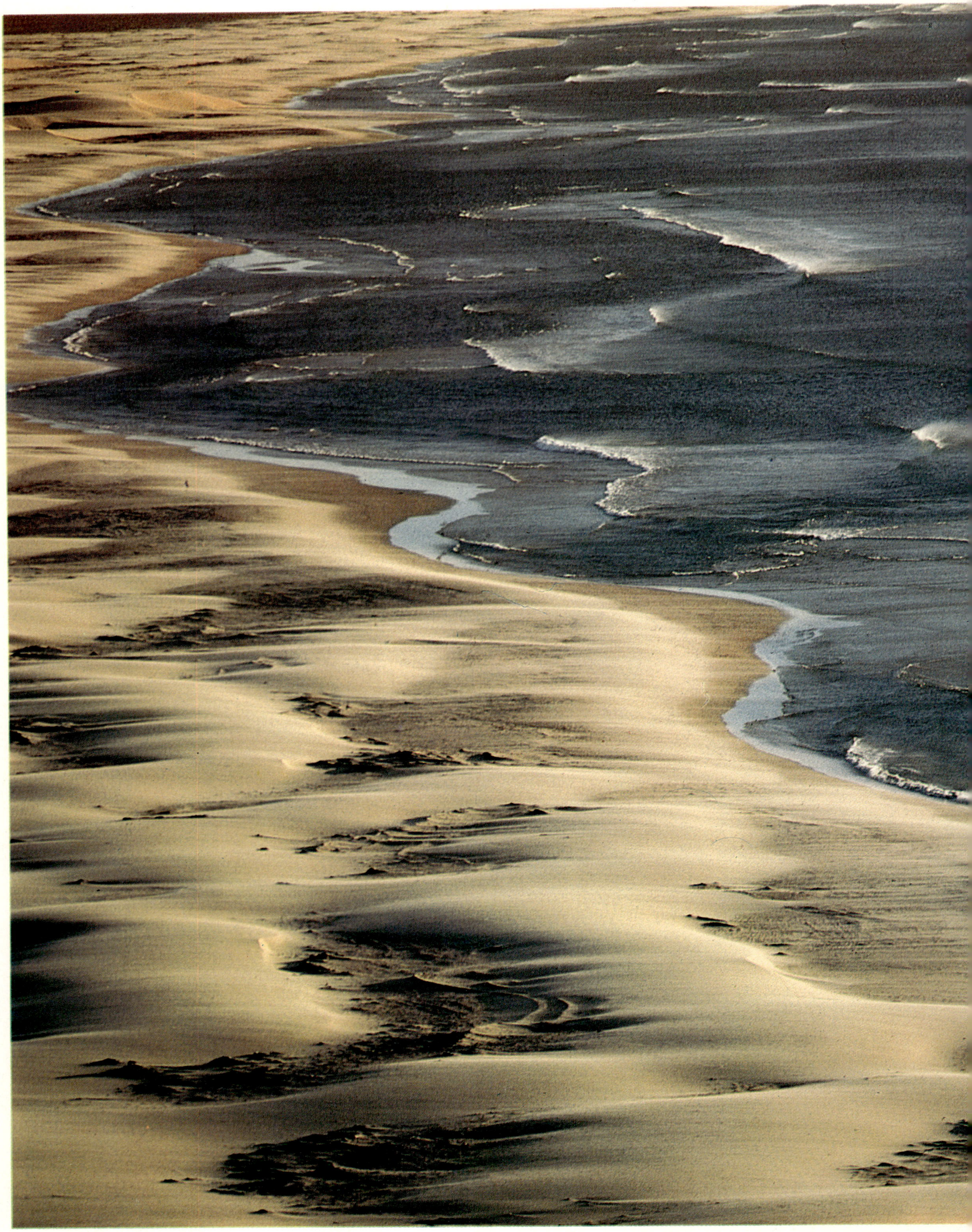

Chapter Six

AT THE OCEAN'S EDGE

No book on the oceans would be complete without a glimpse of their shores. The rim of the sea is not fixed. Winds, waves, currents and rivers are always destroying or building land. Sea cliffs, and beaches of rock, shingle, sand or mud show how the land and sea affect each other.

Each type of shore supports its own set of living things – for instance, plants and animals built for clinging to rocks or crouching in crevices, and creatures that burrow in sand and mud. Seashore animals have adapted in many ways to a home submerged by water at high tide, and left high and dry at low tide. Many island animals, cut off from competitors on the mainland, have evolved in a unique way.

The seashore is rich in plant and animal life and natural beauty. It also provides us with food, building materials and areas for recreation, and serves, too, as a springboard for travel and trade.

A vast sandy beach in South Africa, swept by wind and sea. Sands that build one such stretch of shore may be the crushed, transported ruins of another, for shoreline is always being won and lost in the unceasing war between the sea and the land.

Sculpting the Shores

The seashore is a battleground between the land and the water. In places the sea eats into the land. Elsewhere land thrusts out into the sea.

The Sea Attacks

The sea's main weapons of attack are water, air and stones. Water dissolves limestone rock. Tidal currents press hard on certain shores; but gale-lashed waves are the main erosive agents.

Waves crashing on cliffs compress air into cracks. As waves retreat, this air expands explosively. The cracks grow wider. Lumps of rock break off into the churning sea, which batters them against each other and against the cliff foot, undermining it, until the top collapses. Bit by bit, the sea advances, until all that is left of the cliff-front is a submerged, wave-cut rock platform.

Cliff types vary with the rocks involved. Layered rocks tilting seawards yield steeper cliffs than landward-tilting rocks. Hard rocks like limestone and granite have steeper slopes than soft gravels, clays and sands. Hard rocks also wear away more slowly than soft rocks. But members of both groups hold cracks that lay them open to attack. Widened and deepened by the sea, some cracks give rise to caves and stacks. Local differences between rocks and their resistance to erosion may result in coasts where bays and headlands alternate.

Even where the land does not end abruptly in tall cliffs the sea may be advancing. A shore with boulders and wave-worn hollows shows that the sea is on the offensive.

The Sea in Retreat

The sea builds up land as well as destroying it. Rubble broken from a cliff may go to form a beach elsewhere. Waves and tidal currents are the main removal agents. Wind-driven waves tend to strike land slantingly, so that the SWASH drives sand and pebbles obliquely ashore. But the force of gravity creates a backwash that tugs them straight back down into the sea. Sand and shingle may zig-zag scores of kilometres along a coast by this process known as LONGSHORE DRIFT. But where groynes or headlands bar its path or break the force of waves and currents, the moving debris piles up to form a BEACH.

SANDY BEA
A beach of ro fragments fir ground by th

DUNES
Hillocks of wind-blown sand

BAR
A sand and shingle ridge across a river mouth or bay

SPIT
A slim, low tongue of sand and shingle joined to the land at one end

Beaches

Some sand or shingle beaches are brief strips in coves between headlands. Others form spits, and offshore bars that trap water and form lagoons.

Shingle beaches consist of pebbles – smoothed stones, varying in size between a marble and a hen's egg. Storms drastically change these beaches: scouring chunks away or altering the steepness of the slope.

Many sand beaches consist of grains of quartz derived from sandstone or granite rock. Others are largely built of crushed particles of shell and coral. At low tide on gently shelving shores, some sandy beaches stretch out to sea for several kilometres. Behind them, onshore winds may pile sand into dunes.

New land is also made by rivers bearing tiny particles of clay and silt. Where rivers reach the sea they shed this load. If the sea is shallow and sheltered, the load sinks and piles up until it peeps above the sea as mudflats. Plants root in the mud, creating mangrove swamp or salt marsh. In time, estuary mudflats and deltas may build low-lying land far out into the sea.

HEADLAND A cliff of resistant rock jutting into the sea

STACK A headland cut off when a rock arch fell

BLOWHOLE A hole in a cave roof through which waves hurl spray

SEA CAVE A hole worn by waves attacking a weakness in layers of rock

BAY A dent in the coast made by sea attacking easily eroded rocks

SHINGLE BEACH Pebbles smoothed and heaped up by the sea

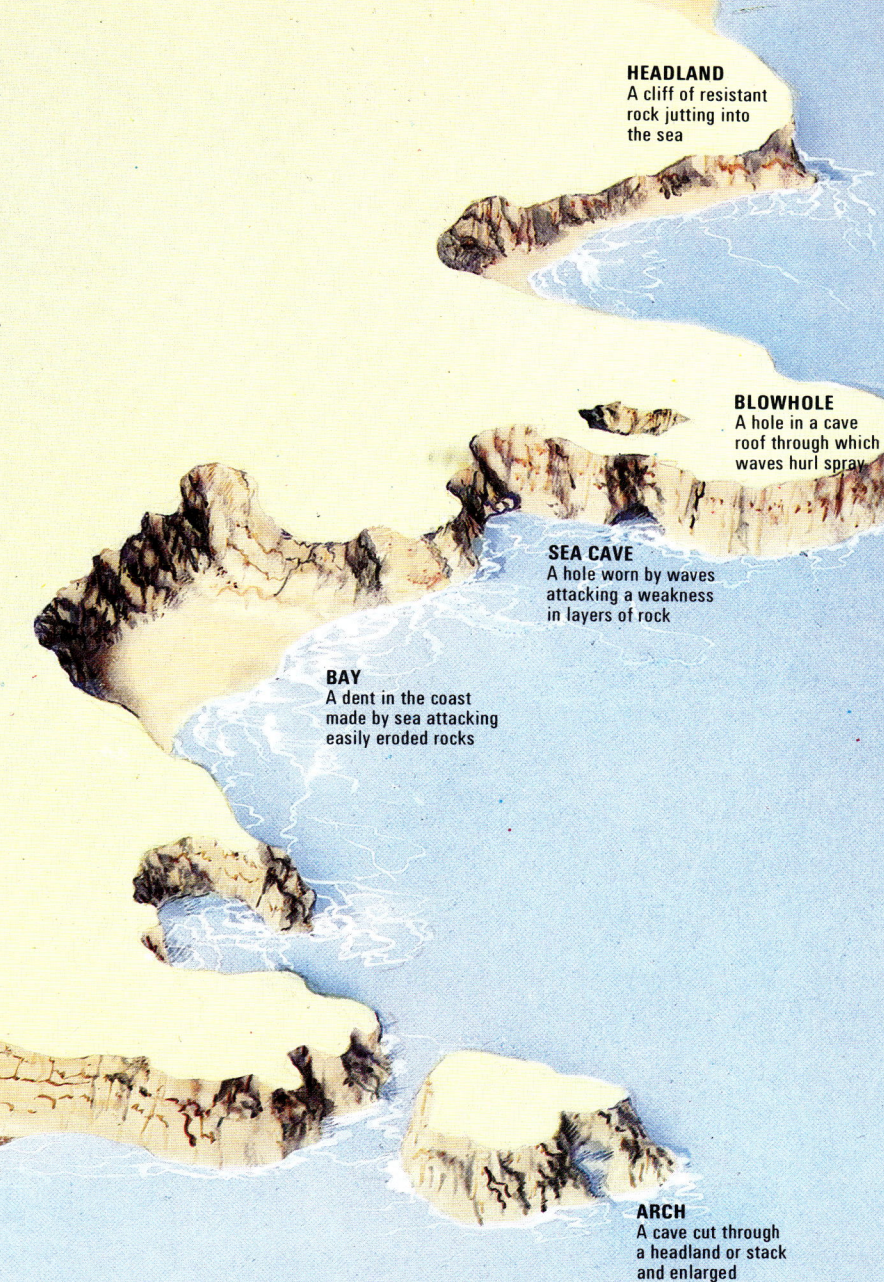

ARCH A cave cut through a headland or stack and enlarged

Waves from the Atlantic smash against the cliffs in South Morocco. The sea is the main agent sculpting the seashores.

Life on the Shore

Each kind of coast has plants and animals which have adapted to its special opportunities and problems. For instance, cliffs above a shore provide gulls, gannets and other seabirds with roosting and nesting ledges safe from predators.

The shore itself supports many animals, which have specifically adapted to cope with its particular problems: the dry periods between high tides, the offshore winds, storm battering and sharp changes in temperature which can occur even on tideless shores.

Animals of Rocky Shores

On rocky shores, limpets and barnacles living high and often dry respire some atmospheric air, while tropical ghost and hermit crabs even have a form of lung as well as gills. To avoid drying out at low tide, sea worms, shore fishes and shore crabs creep under shady stones or hide in pools; sea anemones produce a sticky substance that keeps them damp.

Thick shells or 'perspiration' mechanisms guard certain animals from heat. Some move offshore to escape cold winter weather. Soft-bodied hydroids 'rooted' on rocks survive storms by bending with the waves, while barnacles and limpets are protected by shells.

Animals of Sand and Mud

On sandy shores a different set of adaptations has evolved to meet other opportunities and problems. Here there are no hard, fixed surfaces to cling to or hide under – only a shifting mass of stony grains. But though the surface chills and bakes, dries out and shifts as storm waves batter it, conditions are more settled just a few centimetres down. Moreover any animal beneath the surface remains invisible to predators. Thus the creatures of sandy shores are largely burrowers.

Some molluscs and tubeworms never leave the safety of their burrows, but suck or pull down food into their lairs. However, when the tide swirls in, buried sea worms, bivalves and ECHINODERMS thrust up their feeding tubes or tentacles while an army of crustaceans climbs out to hunt and scavenge. From deeper water, fish move in to snap up the unwary.

Muddy shores also have their burrowers. Mud is rich in food particles but tends to clog breathing organs and, not far down, is poisoned by bacteria. Then, too, estuaries suffer swift changes in salinity and temperature. Breathing tubes and filters, and an ability to withstand a wide range of salinity enable many creatures to live in the mud. Ragworms, *Hydrobia* snails, AMPHIPODS and mysids, for example, teem in northern mudflats.

Nourished by seaweed or plankton swept in on the tide, seashore creatures form a living larder raided from both sea and land. Fishes, birds and man move in to plunder rocky, sandy and muddy shores. Only ice-scoured coasts and shifting shingle beaches are too poor to be extensively exploited by plants or animals.

Above: A crowded clifftop gannet nursery. Sea cliffs form safe bases from which fish-eating seabirds can plummet onto their prey in offshore waters.

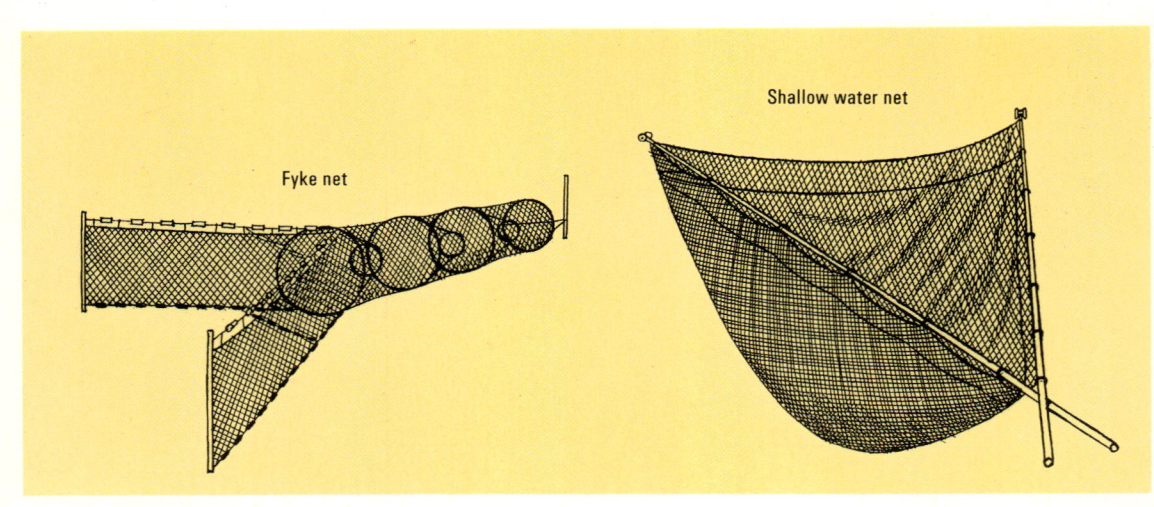

Below left: Two ways of netting inshore fish. The fyke net's linked cones trap fish that swim in with prevailing currents.
Below: Casting a net in west Bali.

ZONES ON THE SEASHORE

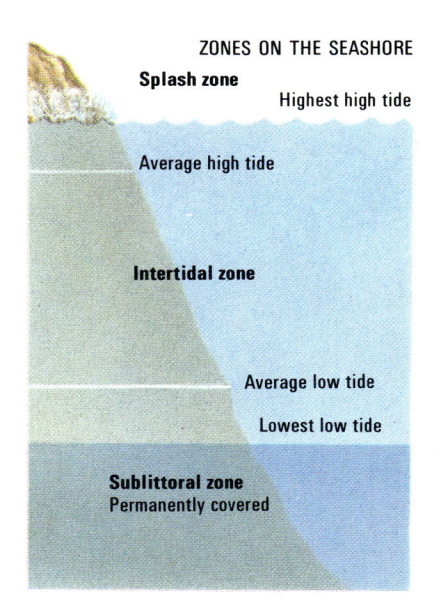

Splash zone — Highest high tide
Average high tide
Intertidal zone
Average low tide
Lowest low tide
Sublittoral zone — Permanently covered

Above: Life zones on a rocky beach. Organisms that readily die in air live below the low-tide level. Ones that stand short spells of drought live in the intertidal zone. Those able to survive an occasional soaking from spray live in the splash zone.

Right: Examples of animals that live in pools on rocky shores. They are dahlia and beadlet sea anemones; common starfish; sea lemon (a sea slug); and eggs of a sea urchin.

Right: Some animals that hide in burrows dug in sandy beaches. At high tide, razor shells, tellins and young gapers thrust up siphons with which they suck in water and scraps of food. Sea potatoes and purple heart urchins use tentacles to grope for food on sand. Ragworms swim and crawl for prey. Lugworms get food from the muddy sand they swallow. Masked crabs burrow to escape from predators.

Below right: Three ways of using bait to catch inshore water animals. Lobsters enter baited wicker 'pots'; fish taking bait get hooked or caught by a springy branch trap.

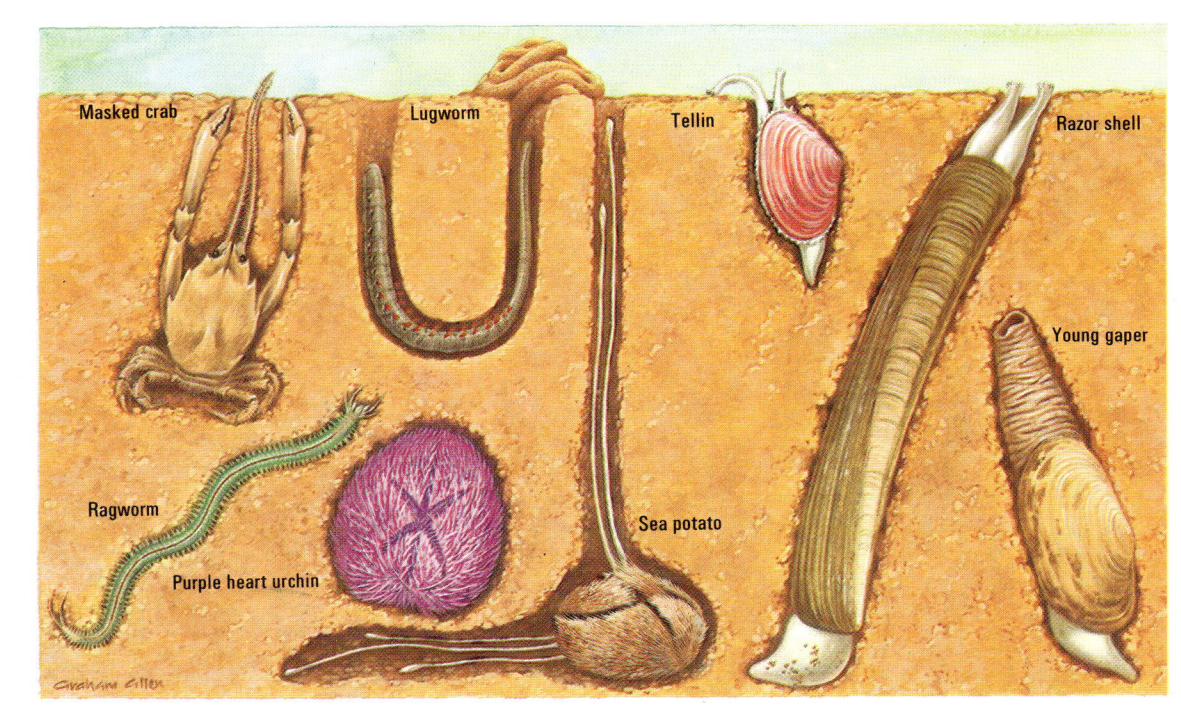

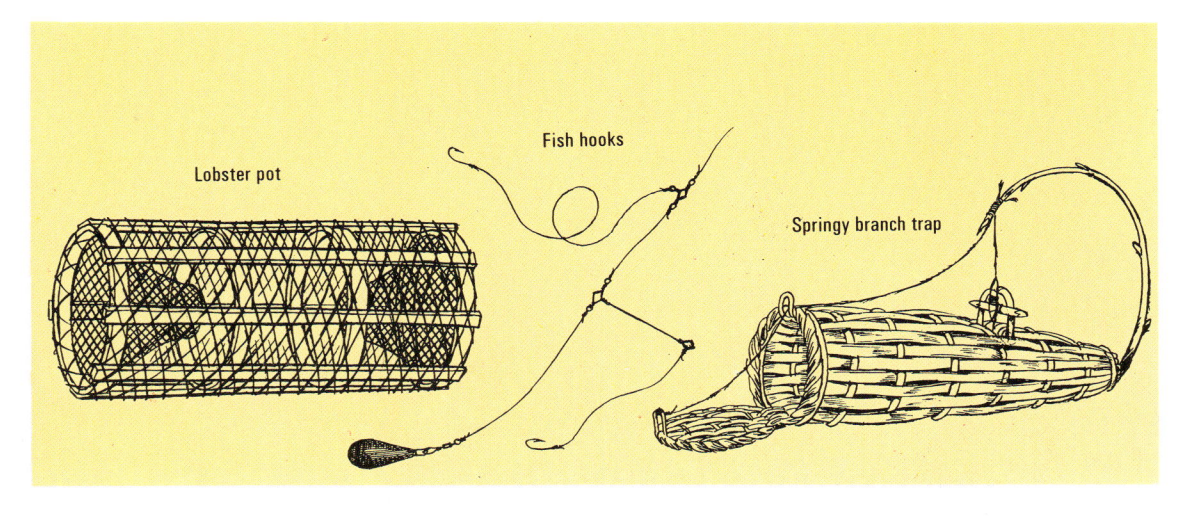

Masked crab · Lugworm · Tellin · Razor shell · Young gaper · Ragworm · Purple heart urchin · Sea potato

Lobster pot · Fish hooks · Springy branch trap

71

Seaweeds

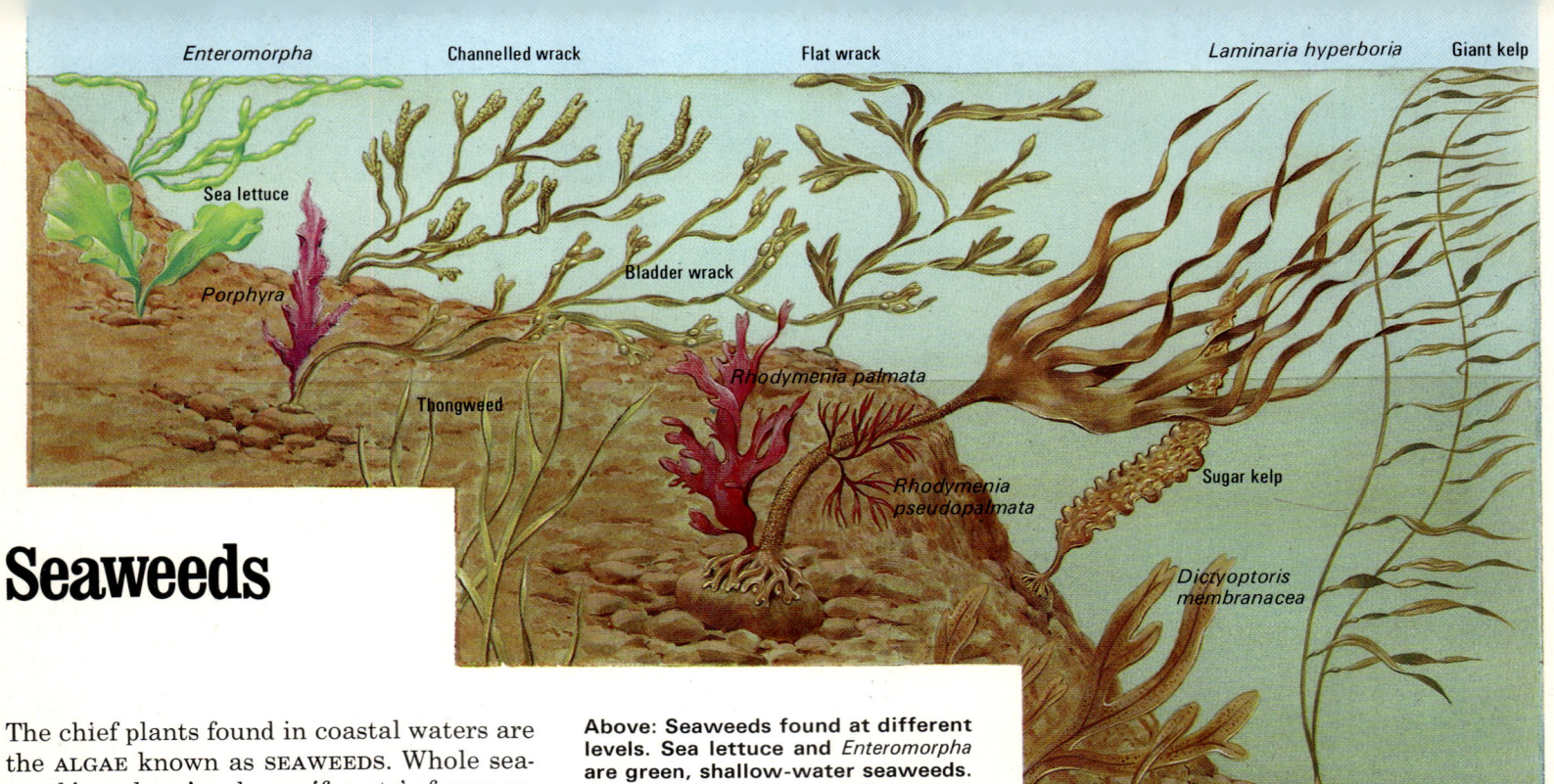

The chief plants found in coastal waters are the ALGAE known as SEAWEEDS. Whole seaweed 'meadows' and even 'forests' of KELP up to 60 metres (200 feet) high grow beneath the waves – especially where rocky shores provide a base. Seaweeds come in a huge variety: stiff, rubbery or flimsy; brown, green, pink, red or purple; skyscraper high or microscopically small.

But all have certain things in common. Unlike most land plants they draw all support and nourishment from water and lack true leaves, stems and roots. Botanists refer to a seaweed's body simply as its THALLUS.

Some species – sea lettuce for example – lack a special thallus shape. Oarweeds grow long, strap-shaped leaflike fronds. But other species have body parts differentiated much like stems and leaves. Bladder wrack has flat 'leaves' buoyed up by pea-sized bladders; coralline seaweeds grow brittle 'stems'; various red seaweeds feature feathery fronds and comblike branchlets.

The rootlike holdfasts gluing seaweeds onto rock, shells, or other seaweeds also differ. Some resemble discs; some claws; and others sticky threads. Holdfasts prevent dislodging by all but the fiercest waves. Tests of holdfast strength have shown that bladder wrack resists forces of up to 42 kilogrammes per square centimetre (600 pounds per square inch).

Seaweeds sprout mainly on the floors of shallow coastal waters because the deeper ocean bed lacks the light they need to photosynthesize. But different types need different kinds of light. Green seaweeds use red light, which only penetrates the sea's top layer. Green seaweeds thrive best on the upper shore. Red light is less vital for the brown seaweeds: these dominate the middle shore. Red seaweeds (including the crimsons, pinks and purples) use blue light, which has deep penetration. Red seaweeds thrive 100 metres (330 feet) down in the warm clear waters of the world.

'Birth' and Death

Seaweeds start life in several ways: as runners; as plantlets sprouting from a frond; and as minute spores. They grow fast – one species up to 30 metres (100 feet) in a year.

But seaweeds have many enemies. Storms annually destroy most wracks off northern shores. Limpets and topshells graze other plants to death. Man, too, in places takes his toll – especially in Japan, where 'seaweed farmers' grow beds of *Porphyra*, a red seaweed highly valued as a food. AGAR, used in medical research to make a culture for bacteria, comes from the red seaweed *Gelidium*. Ice-cream, salad dressings and aspirin are all made with the aid of alginates derived from kelps. Seaweeds once also supplied soda and potash: ingredients for making soap and glass. Iodine also comes from seaweed.

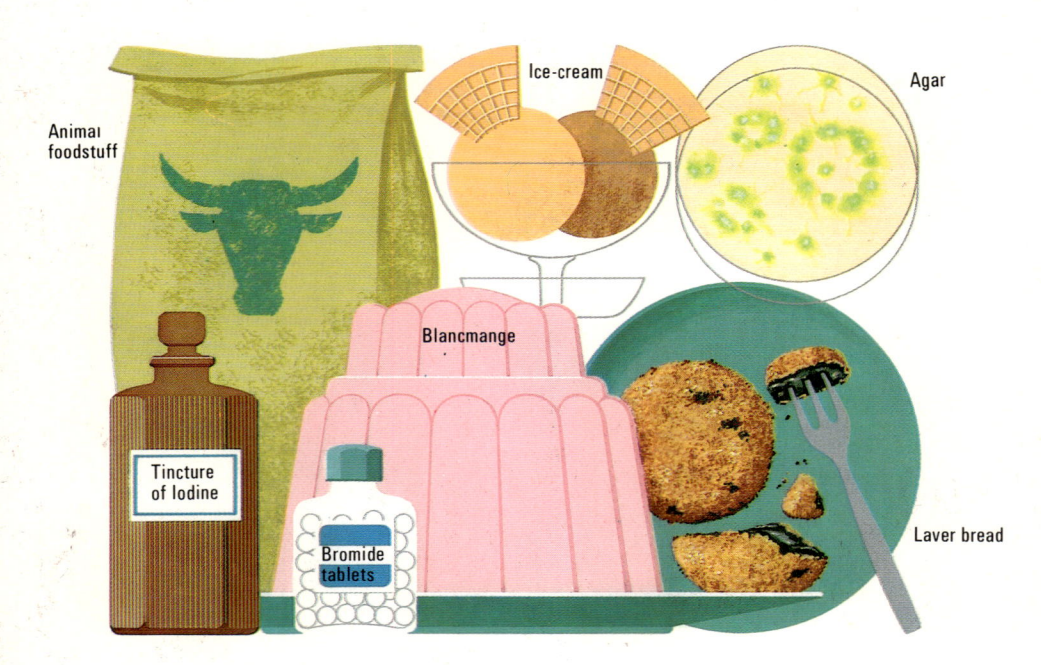

Above: Seaweeds found at different levels. Sea lettuce and *Enteromorpha* are green, shallow-water seaweeds. Channelled, bladder and flat wrack are brown seaweeds of the upper and middle shore. Kelp thrives lower down (giant kelp off Antarctic and sub-Antarctic shores). The red seaweed *Porphyra* flourishes at many shoreline levels, but red seaweeds are the ones best adapted for shady, or deep, dim situations.

Far right: Harvesting seaweed growing on nets off Japan, where people eat seaweeds as delicacies. Nets, fences and buoys, geared to seaweed cultivation, occupy great stretches of Japan's shallow Inland Sea.

Below: Some of the products which are made from seaweed or which contain extracts from seaweed. Bromide tablets are given as a sedative.

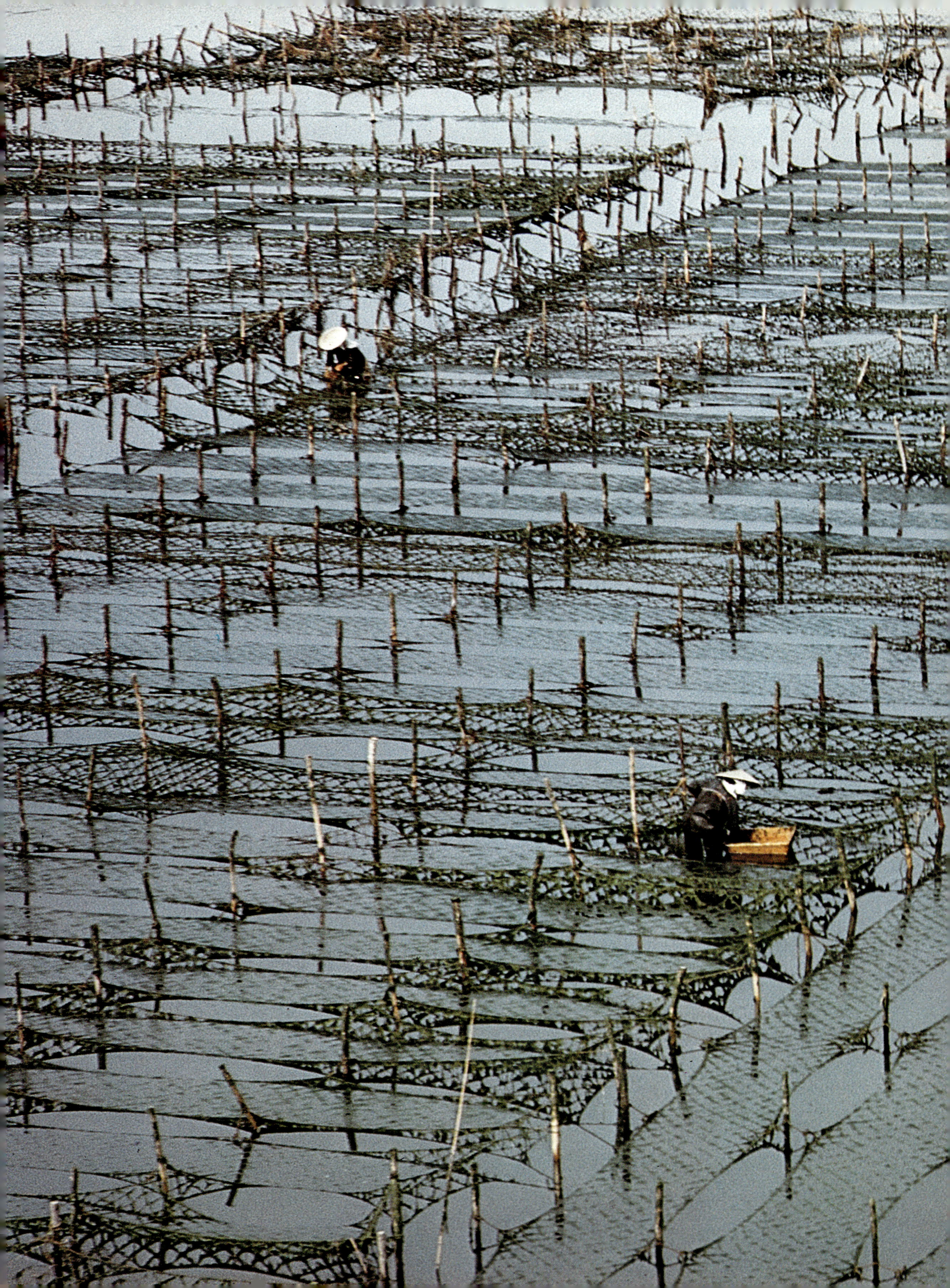

Islands

Left: Von Heron Island on Australia's Great Barrier Reef is a cay or key. Unlike oceanic islands, cays are low, offshore islands formed when the sea heaps coral sand or other debris into sandbanks stable enough for land plants to root in them. The Florida Keys are other well-known examples.

Below: Reef-fringed Raiatea Island in French Polynesia is one of hundreds of Pacific Ocean islands formed when volcanoes thrust up above the sea. Winds and currents brought plants and animals to populate Raiatea's barren, seagirt slopes.

We have already seen that many islands form when chunks of land are cut off from nearby continents. Others, though, appear out in the open ocean when submarine volcanoes thrust up above the surface. Iceland, the Azores, Ascension Island, Mauritius, Tahiti and thousands more originated in this fiery way. Sometimes you can actually see an island born. In 1963 off south-west Iceland the sea boiled and parted to reveal a mound of volcanic cinders about 1·5 kilometres (almost a mile) long that grew into a barren island known as Surtsey. Soon it cooled sufficiently for seeds to take root and grow.

Oceanic islands elsewhere are growing old. The Hawaiian chain has some examples. Here, islands formed one by one at intervals of a million years, probably when molten rock from a 'hot spot' in the mantle punched up through a slowly moving oceanic plate. Tall volcanoes crown Hawaii, the south-easternmost and youngest island. But 2600 kilometres (1600 miles) to the north-west the oldest islands have been worn down, leaving stubs and atolls.

How Living Things Arrive

Islands which formed from land cut off from the continents already had plants and animals living on them. But living things had to cross vast tracts of sea to reach new oceanic islands. Some blew in on the wind. Winds can

Below left: Green turtles and the other four species of marine turtle can reach oceanic islands just by swimming there. These big, aquatic reptiles only haul themselves ashore to lay their eggs, and their young seek safety in the sea soon after hatching.

carry dust-fine orchid seeds, fern spores, insects, mites and spiders thousands of kilometres before they fall – most in the sea, a lucky few on land.

Birds and bats blown off course by winds colonized some islands. Other wildlife drifted in on rafts of vegetation washed down to the sea by mainland rivers, and borne by currents out into the open ocean.

Creatures evolved new forms and habits as they adapted to their isolated island homes. Free from predators, some birds and insects no longer needed to fly, and lost the use of their wings. Without large mammal enemies or rivals, giant tortoises evolved in the Galápagos and Seychelles islands. New Zealand's giant moa and Madagascar's *Aepyornis* (now both extinct) were probably the largest ever birds on Earth.

Exploiting food supplies untapped by land animals, huge robber crabs thrive ashore, and actually drown if kept under water for long. On the other hand, marine iguanas are lizards that swim under water and browse on seaweed off Galápagos shores. In those same islands, the ancestor of Darwin's finches gave rise to 13 species, each designed to tap a different type of food supply.

Unfortunately, when people came to live on the islands, some of the creatures already there could not compete with the animals they brought. Many species died out.

Below: Marine iguana feeding on seaweed off a Galapagos shore. These unique seaweed-browsing lizards may derive from one or two females that drifted in on vegetation washed out to sea from the River Guayaquil in Ecuador.

Above: Kiwis are flightless birds the size of chickens. They live only in New Zealand. There, too, once lived their kin the giant moa – a bird twice as tall as a man. Flightless birds, some unusually large, evolved on remote islands where the sea kept them safe from predators.

Large ground finch

Warbler finch

Woodpecker finch

Evolution has fitted Darwin's finches to fill feeding niches held elsewhere by other birds. The large ground finch's big beak cracks seeds. The warbler finch's fine bill takes winged insects. The woodpecker finch has learnt to use twigs to tweak grubs from holes in rotten wood.

Using the Shore

Man exploits the sea's edge in many ways – for food; as a base for travel and communication; for recreation, and as a reservoir and quarry.

Ports and Harbours

As world trade and military seapower expanded, maritime nations built more and larger harbours and ports. Harbours are simply sheltered anchorages; ports are harbours with aids for loading and unloading.

Kiel, San Francisco, and Scapa Flow are all examples of fine natural harbours. Southampton and Portsmouth owe much to the water protected by the Isle of Wight. Liverpool, Le Havre and New York all grew up on sheltered river mouths.

Most major harbours are also ports. New York, one of the largest in area, sprawls over a waterfront of 1215 kilometres (755 miles) and has enough berths to handle 391 vessels simultaneously. Ports are often artificially protected and kept open. The cross-channel port of Dover and many others rely on breakwaters to break the force of waves. Then, too, dredgers may deepen ports to open them to large, fully laden vessels. (Supertankers need such deep water that they are usually loaded and unloaded at special offshore terminals.) Frequent dredging may be required for ports in estuaries which tend to silt up.

Ready access to foreign raw materials and

Above: Mining sands in Australia. Both quarrying and mining are fairly common along the shore.

Left: An oil refinery near Port of Spain in Trinidad. Many industries grow up close to the coast, in easy reach of ports and shipping transport.

Below: The port of Barcelona in eastern Spain. Ports are important centres for imports and exports. As world trade has increased, ports have become larger and more complex.

Below: Dounreay atomic power station on the north coast of Scotland. Such power stations are usually situated near the sea where they can get the vast quantities of water needed for absorbing the waste heat that they produce.

markets has encouraged industries to spring up on, or near, big ports. Seaside power stations use seawater to soak up the huge amounts of waste heat.

Planning and building modern ports is complex and costly. First, surveyors study local tides, wave heights and directions, seabed depths, channels, sand bars, silting rates and other factors. Then, builders install lighthouses, channel markers and other navigation aids, breakwaters, berths, turning basins and railway sidings.

Resorts

Since the late 1700s, when sea-bathing began to be a popular form of recreation, much coastal building has been to cater for the holiday trade. Countless resorts, from Benidorm to Barbados, have sprung up on once empty beaches to house and feed the millions who come to enjoy themselves beside the sea.

Here, particularly, people build groynes to guard beaches against erosion caused by longshore drift. Coasts heavily attacked by storm waves may be reinforced by sea-walls – low, artificial cliffs resisting pressures of up to 1,660,000 kilogrammes per square metre (340,000 pounds per square foot).

Quarrying the Shore

People not only build on the coast, they also cart parts of it away. Limestone sea cliffs in southern England yield Portland stone. Elsewhere the building industry strips beaches of huge quantities of sand and shingle. The shore is modified by the many ways in which it is used.

Right: St Malo beach in Southern France. Swimming, sailing, water skiing and sunbathing are just a few of the ways that people use the seashore for pleasure.

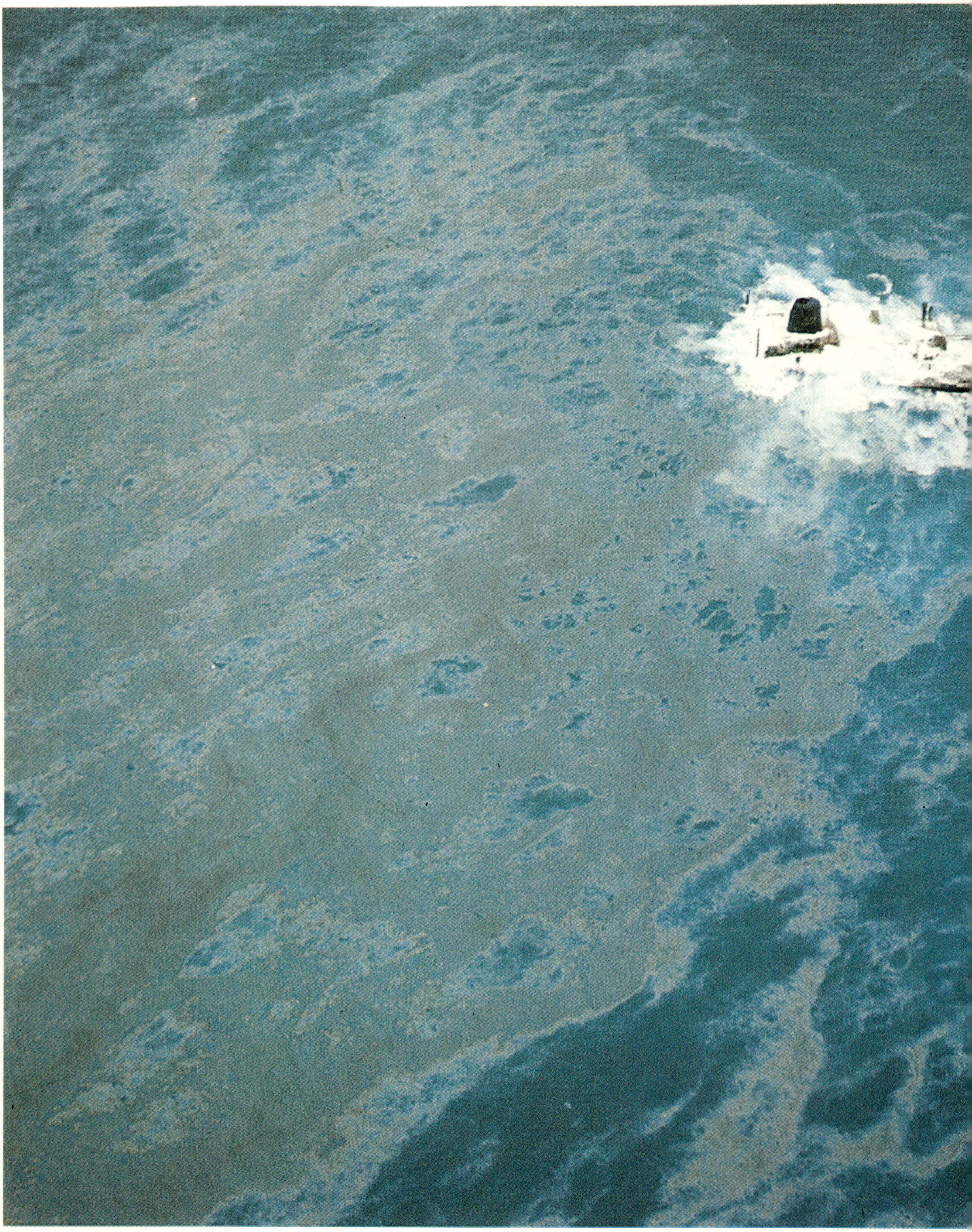

Chapter Seven

TOMORROW'S OCEANS

Until quite recently, people had little real effect upon the oceans. Much of what we are doing now, however, is damaging them. Lethal wastes pour hourly into seas from cities, factories, farms and ships. Some poisons take years to build up into deadly quantities, and centuries to spread throughout the oceans. People have tended to overestimate the capacity of the sea to handle pollutants; only action taken now to curb pollution may save tomorrow's waters.

There are also plans to manipulate the oceans with schemes to dam a sea, to wall off oceans from each other, or to use heavy hydrogen from seawater to harness the immensely powerful kinds of energy that heat the Sun. Less grandiosely, people are already planning to build ports and cities on artificial offshore islands.

Whether or not these ideas actually materialize, the future of the oceans – and thus of ourselves – rests largely on how far nations will cooperate to keep the waters clean, and how far they will be willing to share their wealth.

Thirty thousand tonnes of crude oil spilt out from the tanker *Torrey Canyon* when she sank off south-west England in 1967. Since then far greater oil leaks have contaminated coastal seas. But oil is only one of many noxious substances now threatening the health of ocean waters.

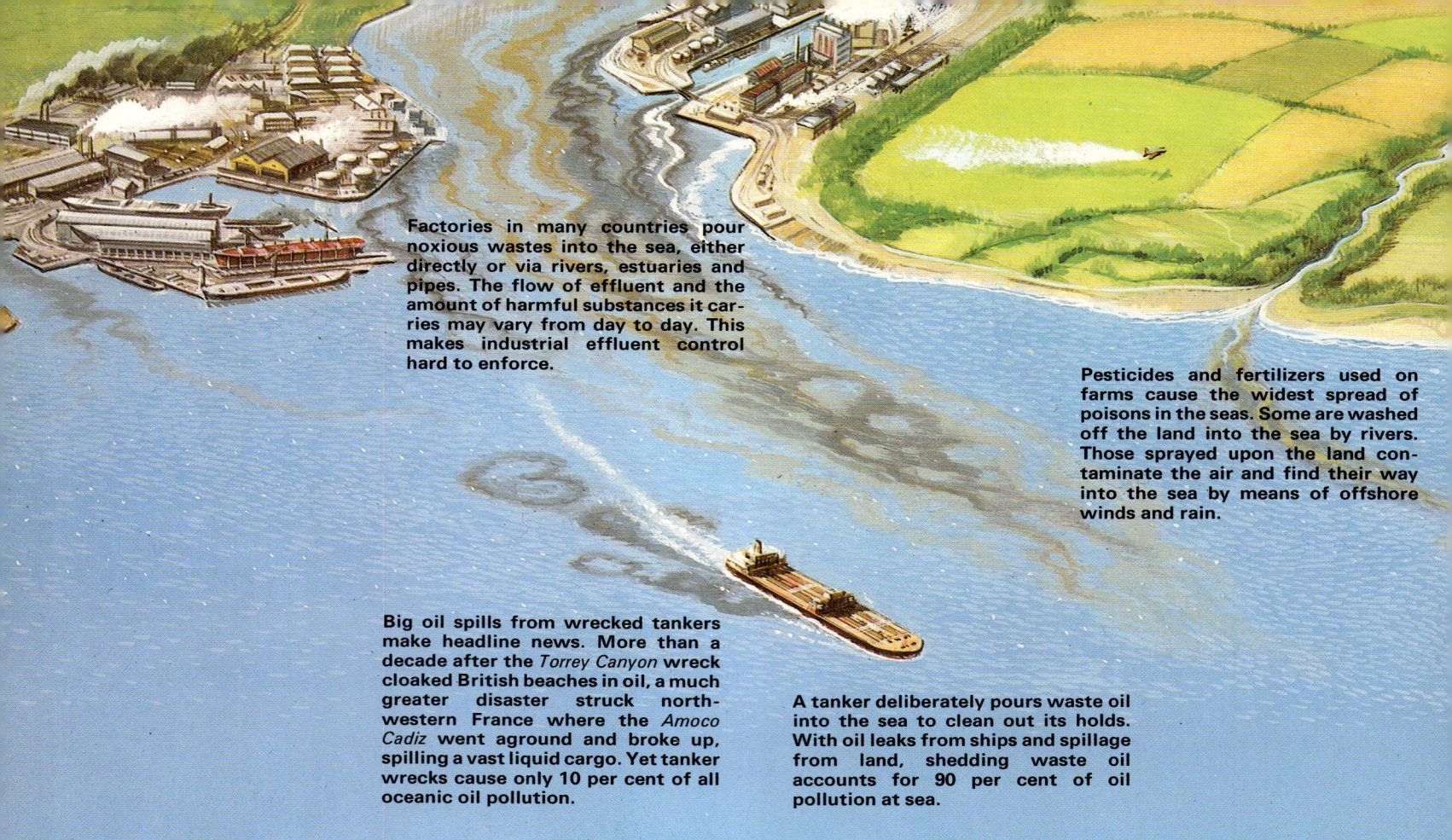

Factories in many countries pour noxious wastes into the sea, either directly or via rivers, estuaries and pipes. The flow of effluent and the amount of harmful substances it carries may vary from day to day. This makes industrial effluent control hard to enforce.

Pesticides and fertilizers used on farms cause the widest spread of poisons in the seas. Some are washed off the land into the sea by rivers. Those sprayed upon the land contaminate the air and find their way into the sea by means of offshore winds and rain.

Big oil spills from wrecked tankers make headline news. More than a decade after the *Torrey Canyon* wreck cloaked British beaches in oil, a much greater disaster struck north-western France where the *Amoco Cadiz* went aground and broke up, spilling a vast liquid cargo. Yet tanker wrecks cause only 10 per cent of all oceanic oil pollution.

A tanker deliberately pours waste oil into the sea to clean out its holds. With oil leaks from ships and spillage from land, shedding waste oil accounts for 90 per cent of oil pollution at sea.

Endangered Oceans

These scenes reveal some of the ways in which people carelessly and accidentally damage the waters of the oceans.

Chapter Three showed how over-fishing of the oceans threatens future fish supplies. But people not only take too much of what they want from the ocean: they also use it as a dumping ground for too much waste.

Much of the damage done by thoughtless waste disposal is already visible. But the full effects of present folly may not show up for many years – certain substances take a long while to build up into concentrations sufficiently toxic to kill marine plants and animals, or to foul the shores and surface waters of the seas.

Poisoning the Waters

Some coasts and shallow seas are already polluted by sewage piped from cities and resorts. Fertilized by human wastes, floating algae multiply explosively, and the rotting mass of plants remove vital oxygen from the upper waters where plants normally thrive. Few living things can survive in waters thus damaged by EUTROPHICATION.

Oil is another serious pollutant. Some 6 million tonnes are annually spilt into the sea by tankers, or from offshore oil fields. Like an even larger amount of solid litter dropped into the sea, oil spills disfigure beaches. But they also do much graver damage. The *Torrey Canyon* and *Amoco Cadiz* shipwrecks, and the Santa Barbara oil leak, to name just three disasters, caused the death of innumerable underwater organisms. Millions of seabirds too were suffocated and drowned, or chilled to death when the oil clogged their feathers and spoilt their natural insulation. The increasing number of offshore oil rigs multiply the risks of future major spills.

Many chemicals discharged into the sea have dangerous effects. In the 1950s and 1960s, factories poured mercury waste into Japan's Minamata Bay. The mercury accumulated in fish and shellfish eaten by the local people, scores of whom suffered brain damage, blindness, loss of muscle power, or death. Today, fishes' bodies are still accumulating DDT and other pesticides washed out of the atmosphere or off the land into the sea by rivers years ago – before we knew that birds and people who ate fish contaminated with DDT could dangerously concentrate these poisons in their own bodies.

Radioactive particles are yet another hazard. No one knows for sure what damage may yet result from past nuclear bomb tests in and over oceans, and from the wastes leaked by

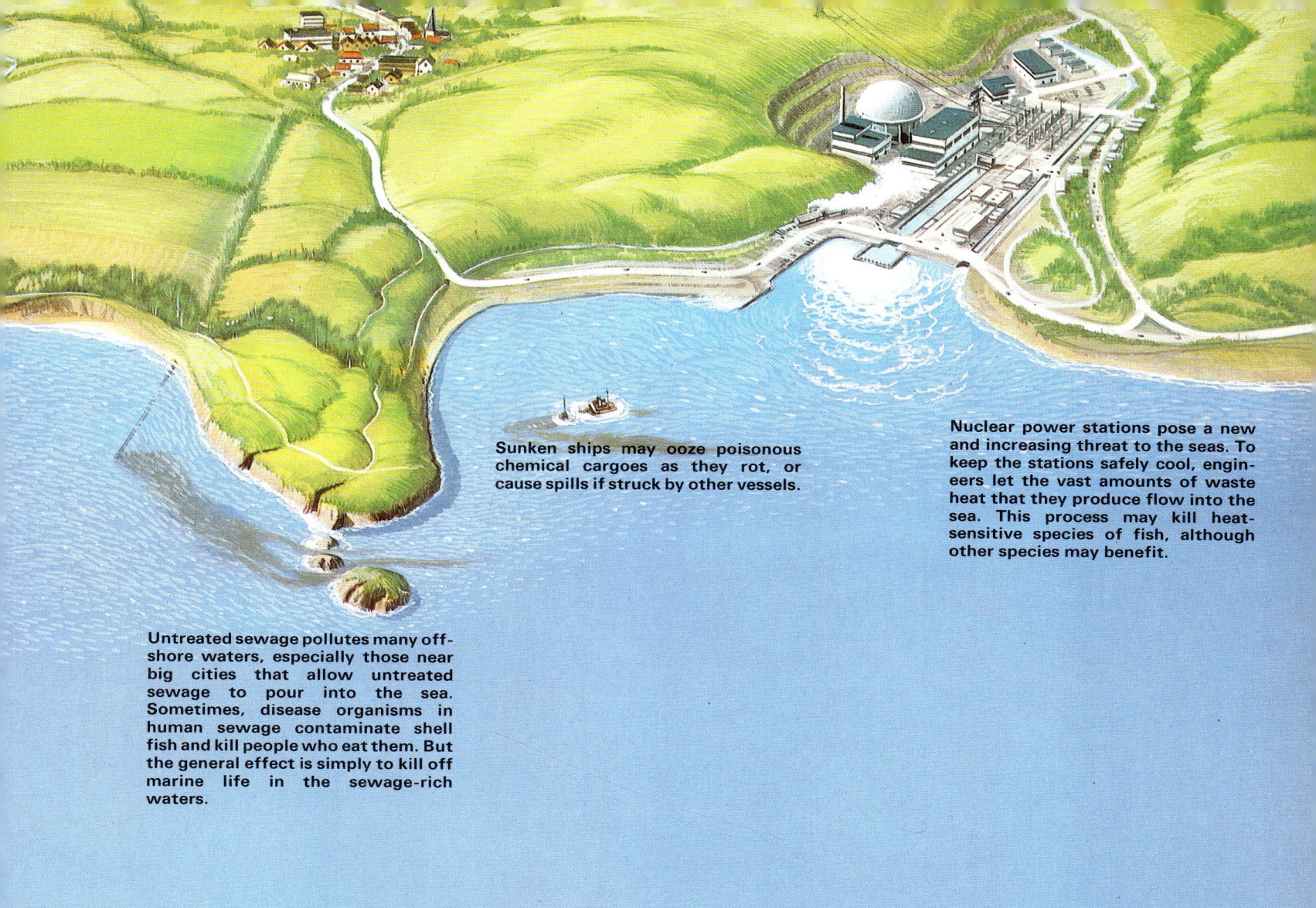

Sunken ships may ooze poisonous chemical cargoes as they rot, or cause spills if struck by other vessels.

Nuclear power stations pose a new and increasing threat to the seas. To keep the stations safely cool, engineers let the vast amounts of waste heat that they produce flow into the sea. This process may kill heat-sensitive species of fish, although other species may benefit.

Untreated sewage pollutes many offshore waters, especially those near big cities that allow untreated sewage to pour into the sea. Sometimes, disease organisms in human sewage contaminate shell fish and kill people who eat them. But the general effect is simply to kill off marine life in the sewage-rich waters.

In places, poisonous chemicals have been dumped on the seabed in metal containers. If the metal corrodes, the poisons escape and are spread by deep-sea currents.

A second danger comes from the radioactive wastes created by nuclear fission. No one can be sure of the long-term effects of such wastes that are dumped in thick concrete casing on the deep ocean floor.

nuclear power stations now springing up on shores around the world.

POLLUTION from various sources is locally already bad and likely to get worse. Many bays and the almost enclosed Baltic and Mediterranean seas are now heavily affected. Conservationists argue that these waters will end as stinking sinks – unless we stop pollution now.

Saving Tomorrow's Seas
Pollution could be curbed in several ways. For instance, land-based purification of sewage and industrial wastes prevents poisons flowing out into the sea. Special safeguards in loading and cleaning oil tankers' tanks reduce oil spills. Keeping to shipping lanes cuts down the risk of collision.

Since 1968, the Group of Experts on the Scientific Aspects of Marine Pollution (GESAMP) has been working out how best to clean the seas. The answer plainly lies in international anti-dumping laws. The United Nations Intergovernmental Consultative Organization (IMCO) has held several conventions, but countries have proved slow in backing it. By the end of the 1970s, the future health of the oceans was still at risk.

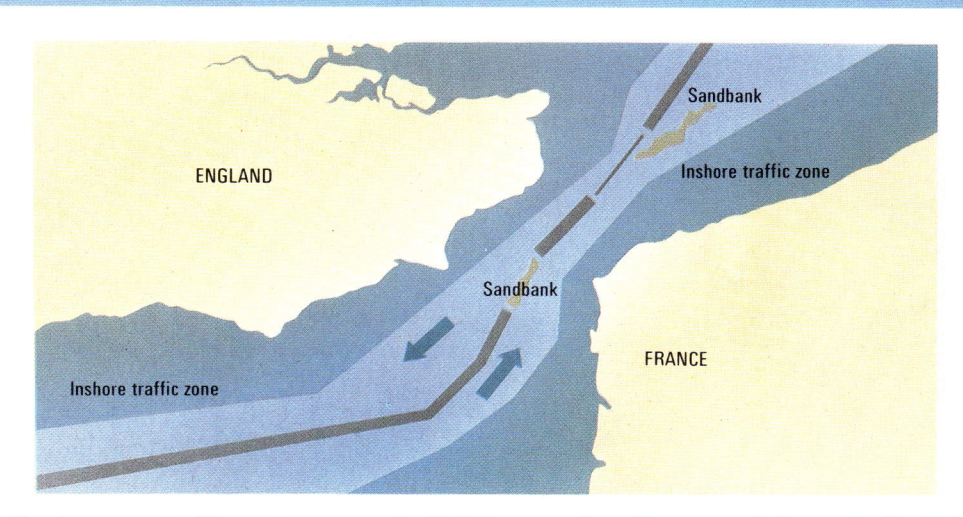

The two-way traffic system set up in 1967 for vessels sailing up and down the English Channel, past the Straits of Dover. Mid-Channel sandbanks in places form a natural barrier to big, deeply laden craft. Long enforced for land and airborne traffic, such schemes are only now becoming used for busy seaways. Properly enforced, they could reduce collisions and the resulting spillages that cause pollution.

81

Schemes for the Future

People have already learned how to reclaim land from the sea, and may soon be tapping the energy of the waves, turning icebergs into floating reservoirs, and mining the floor of the abyss. Even stranger possibilities could lie ahead.

Man-Made Islands

In theory, ports and cities could be built out in the sea or on its bed. Dutch companies headed by the Bos Kalis Westminster Dredging Group have shown how dredging ships could help build gravel islands in the North Sea. The islands could serve as bases on which to build airports, oil terminals, ports, or waste disposal plants. Britain's Pilkington Brothers have planned a floating Sea City sited off East Anglia. Offshore power stations and factories on man-made islands, underwater storage tanks, and even underwater cities have been proposed by land-hungry Japan.

Walls Across the Sea

The grandest schemes affect whole seas and oceans. Thus in 1928 Germany's Herman Sörgel proposed Project Atlantropa. Dams with hydro-electric power plants would block each end of the Mediterranean. Robbed of much inflowing water, the sea would shrink and yield huge tracts of dry land.

Equally ambitious is a proposed dam 34 kilometres (21 miles) long across the Strait of Dover. This scheme would provide a bridge from France to England, and produce great quantities of energy.

Both schemes are dwarfed by Pyotr Borisov's plan to dam the Bering Strait between the Soviet Union and Alaska. Pumps in the 72 kilometre (45 mile) long dam would drive cold water from the Arctic Ocean to the Pacific. Warm Gulf Stream water would then flow into the Arctic Ocean from the Atlantic, heating the cold northern rims of Asia and North America.

All these schemes have snags. Sheer cost is one. Disruption of established human settlements and trade is another. But the unknown effects upon world climate could prove most serious of all. Displacing Gulf Stream water to melt Arctic ice might actually trigger off a new cold phase in the present Ice Age. To stop this redirected flow occurring naturally, one American engineer proposed the wildest scheme of all: a wall across the North Atlantic.

Right: Towing an iceberg from a shipping lane. Towed to hot, dry lands, many icebergs, each several square kilometres in area, may serve as vast freshwater reservoirs. Nine-tenths of a plastic-wrapped Antarctic berg could survive a ten-month haul to southern California.

Energy from Heavy Hydrogen

Less fanciful than any of these schemes is the possibility of using an ingredient in seawater to supply most of the world's energy requirements. This substance is deuterium or 'heavy hydrogen'. It occurs in seawater as one atom for every 3000 ordinary hydrogen atoms. (Water, of course, consists of hydrogen and oxygen.) At very high temperatures, the nuclei of deuterium atoms fuse, producing helium and vast amounts of energy. Uncontrolled nuclear fission supplies the power packed by hydrogen bombs. Controlled fusion could yield enough useful energy to fuel our world for several thousand million years. The problems of building nuclear fusion reactors are huge, but scientists may solve them before the end of this century.

Sharing the Seas

Tomorrow's world may also see international laws to settle questions like who owns which straits, fisheries, and seabed mineral deposits. By the late 1970s United Nations conferences had reached no full system of agreements. But as the importance of the oceans is becoming more and more apparent, cooperation on their exploitation becomes more and more vital so that inland countries and developing countries can benefit from them as well.

Left: A model of Sea City, designed by the British firm of Pilkington Bros. Years of study went into this scheme for a city capable of being built in the North Sea off the coast of eastern England. Outside the city proper, breakwaters would fend off storm waves, and ships could load and unload at quays equipped with cranes. The city area itself would feature sheltered basins and facilities for work and recreation.

Below: Gaseous plasma glows purple in an experiment that could make the heavy hydrogen in seawater a vital source of energy produced by nuclear fusion. Fusion can only happen in plasma at a temperature of many million degrees Centigrade. If the superheated plasma touched the sides of its container these would melt. Experimental devices like the one shown use magnets to make sure this cannot happen.

Projects being developed for the future

Thermal power from the differences in water temperature.

Power from wave energy using Salter's ducks, wave contouring raft, oscillating water column, or Russell rectifier.

Towing icebergs to dry lands from the Antarctic to provide fresh water.

Large cargo submarines travelling under the Arctic on a short route between North America, Western Europe and Japan.

Extensive fish farming.

Exploitation of different species of fish for food, such as the lantern fish, grenadier, director, and black scabbard.

Extraction of substances from manganese nodules.

Mining of seabed deposits by means of remote controlled submersibles.

Development of subsea oil production systems.

Development of man-made islands and floating cities, factories, hotels, etc.

Use of heavy hydrogen to produce energy in nuclear fusion.

Dictionary of Oceans

A

Abyss is the deep ocean floor and all regions deeper than 3700 metres (12,000 feet).

ACMRR (Advisory Committee on Marine Resources Research), established by the Food and Agriculture Organization of the United Nations (FAO), advises international agencies including the IOC on matters relating to the living resources of the oceans.

Agar is a gelatinous substance obtained from certain red ALGAE. The best agar is extracted from species of *Gelidium*, *Pterocladia* and *Gracilaria*. One important use of agar is in growing bacteria for medical and scientific research. Unlike gelatin, which it replaced towards the end of the last century, agar remains undigested by most bacteria. The agar provides a firm clear jelly on which to grow the bacteria and fungi under investigation.

Algae form one of the four divisions in the plant kingdom. Simpler in structure than garden plants, algae are largely aquatic and range in size from a single cell like the DIATOM to large seaweeds such as the KELP. See also BROWN ALGAE, GREEN ALGAE, RED ALGAE.

Alternation of generations occurs in all sexual life cycles. In plants the generation producing the reproductive cells or GAMETES, the *gametophyte*, alternates with that resulting from the combination of male and female gametes following fertilization, the *sporophyte*. In most higher plants, the sporophyte is the dominant generation. Among ALGAE, however, many species have a life cycle dominated by the gametophyte generation. Among members of the animal kingdom, the gametes (eggs and sperm) produced by sexually mature animals are usually short-lived, but there may be more than one generation in the life history. For example, in some COELENTERATES a sexually reproducing generation gives rise (by the union of egg and sperm) to an asexual generation that produces further new individuals by budding or simple division of the 'parent' animal.

Amphipods are shrimplike CRUSTACEANS. Many of them are agile jumpers such as sand or beach 'hoppers'.

Anadromous fish such as salmon live in the sea, but migrate into fresh water to breed.

Anodic effect – see ELECTRICAL FISHING

Aphotic zone of the sea is that portion starting from a depth of about 1000 metres (3280 feet) which receives no sunlight.

Aquaculture is the cultivation or propagation of aquatic life including seaweeds, shellfish, crustaceans and fish.

Arrow worms – see CHAETOGNATHS

Atoll is a ring-shaped CORAL reef that encloses a lagoon in which there is no land and which is surrounded by the open sea.

Azoic means devoid of life.

Aquaculture in Southern France: oysters are cultivated on empty mussel shells.

B

Barrel is a unit of measurement often used for oil. One barrel of crude oil equals 42 US gallons (159 litres).

Barrier reef lies offshore from a land mass with a lagoon between. There are usually openings in the reef, particularly where rivers flowing into the lagoon create conditions unsuitable for the CORALS.

Bathyscaphe is a deep diving SUBMERSIBLE. It is an 'underwater balloon' consisting of a pressure-resistant 'gondola' suspended beneath a float. Instead of air a liquid lighter than seawater, usually petrol, is used. Vents in the float permit seawater to enter beneath the petrol and so equalize the pressures inside and outside its thin walls. The bathyscaphe is ballasted with metal to dive. To ascend, the ballast is released.

Beach is the seaward limit of the shore. Its own extent is marked approximately by the highest and lowest water levels.

Beaufort Wind Scale was originally devised by Admiral Beaufort in the 19th century. He classified wind forces by the amount of sail carried by a ship. It has since been modified and expanded for international meteorological use.

Bends – see DECOMPRESSION

Benthic division is a primary division of the sea which includes all of the ocean floor.

Benthos are the plants and animals that are attached to or crawl on the seabed.

Bio-accumulation is the capacity of many living things to accumulate substances of which only traces are found in the surrounding environment. These substances range from pesticides and other pollutants to naturally occurring elements such as vanadium and copper.

Biogenous sediment consists of at least 30 per cent by volume of particles derived from the skeletons of organisms such as DIATOMS (silica) and foraminifera (calcium carbonate).

Bioluminescence is the 'cold' light produced by living things. Often mistakenly called phosphorescence, it is produced by members of most groups of marine organisms including bacteria, plants and animals. Some deep-sea squids, for example, produce a cloud of luminous particles rather than the inky 'smoke screen' released by shallower water species to deter attackers. The great majority of fish species found in deep waters are luminous.

Biomass is the total weight of living matter in a given area or volume.

Bivalves are MOLLUSCS which have two hinged shells. Generally they spend most of their life on the seabed or burrowing into soft sediments, but some species bore into rock or wood. Typical examples are the clams, oysters and mussels. Some of the burrowing forms, however, are not so easily identified as bivalves. The shells of the ship worm *Teredo*, for example, are reduced to two small plates with rasp-like edges which cut away the wood it feeds on. Many of the bivalves that live in sand and mud have a hatchet-shaped foot that they use for burrowing. A few bivalves, for example the Queen scallop *Chlamys opercularis*, swim by flapping their shell valves vigorously.

Bloom describes a high concentration of PLANKTON, usually phytoplankton. The organisms may become so numerous that they discolour the water producing, for example, RED TIDES.

Blow-out preventer is used to stop oil or natural gas 'gushing' from a well. Generally it is permanently attached to the top of a well. The most common type consists of two pistons directly opposite one another which can be closed together either hydraulically or manually. The ends of the pistons, which are made of tough neoprene rubber, make a perfect seal when pushed together.

Bony fish are by far the most common form of fish: every living CARTILAGINOUS FISH (sharks, rays and chimaeras) is matched by more than thirty species of bony fish. Bony fish occupy practically all aquatic niches. Some species, such as the mud-skippers, can even move and breathe on land.

Bore is a tidal phenomenon which occurs when water is forced by tidal action into large but rapidly narrowing river estuaries. The excess water escapes by moving upstream in a wavelike front over the river water.

Boundary currents flow parallel and close to the CONTINENTAL SHELF. They are caused by deflection of the prevailing eastward- and westward-directed CURRENTS by the continental land masses.

Brackish water is water with a salinity of less than 17,000 parts per million and more than 500 parts per million.

Spilling breaker

Plunging breaker

Breakers are waves breaking on the shore. They are roughly classified as spilling breakers, plunging breakers, surging breakers and collapsing breakers.

Breathing mixtures – see NITROGEN NARCOSIS

Brown algae (Phaeophyta) contain a substantial quantity of brown pigment called fucoxanthin which gives them their characteristic colour. The largest and most conspicuous marine algae belong to this group. Certain species of *Macrocystis* and *Neocystis* of the Pacific and Antarctic regions reach lengths in excess of 33 metres (108 feet). Many of these seaweeds have bladders that act as floats to keep them at the surface. Free-floating masses of the gulfweed *Sargassum* are found in the Gulf Stream and Sargasso Sea.

Buoy is a floating object, other than a lightship, moored or anchored to the bottom and used mainly as an aid to navigation. Buoys vary in shape or colour. A buoy with an identifying shape on top is called a topmark buoy. Other devices such as sound, light, radio or sonar signals, or a combination of these, may be used to help locate and identify buoys. In recent years, the large navigational buoy or lanby has begun to replace lightships, even lighthouses. Large buoys are also being increasingly deployed in the open ocean to gather scientific and meteorological information.

C

Carrack was a sailing ship first built in the 13th century. Unlike previous ships its sails consisted of both the triangular LATEEN (pronounced 'latin') sail and the square sail. The carrack set the basic design of the sailing ships which eventually were to conquer the oceans.

The Spanish caravel in which Columbus discovered America. The caravel was a later improved version of the carrack.

Cartilaginous fish have a comparatively soft skeleton reinforced by bony plates. Unlike most BONY FISHES, they lack a swim bladder to maintain their buoyancy: being heavier than water, cartilaginous fish must continue swimming to stay at any particular depth. Cartilaginous fish are carnivores whereas bony fish may be herbivores, omnivores or carnivores. Two other features set them aside from the bony fishes. Unlike most bony fishes, fertilization of the eggs in all cartilaginous fish is internal. Secondly, they maintain the water balance of their body fluids by retaining urea. There are three groups of cartilaginous fishes: sharks, rays and chimaeras, amounting to some 575 species in all. The sharks include the largest species of fish: the whale-shark can reach a length of 20 metres (65 feet). Among the rays, the skates have the greatest number of species, about 100, of cartilaginous fish. The chimaeras, because of their long whip-like tails, are often known as rat-tail fish.

Catadromous fish such as the eel live in fresh water but migrate to the sea to breed.

Cephalopods are bottom-dwelling or swimming MOLLUSCS possessing a large head, large eyes, and a circle of arms or tentacles around the mouth. The class Cephalopoda includes the octopus, squid and nautilus. The cephalopods have attained the largest size of any invertebrates (animals without backbones). One giant squid, *Architeuthis*, found in the North Atlantic measured 20 metres (65 feet) including the tentacles.

Chaetognaths are small, narrow, wormlike animals that are found swimming in the sea from the surface to great depths. Some species, especially *Sagitta*, are used as INDICATOR SPECIES.

Christmas tree is the name given to the collection of valves installed on a production well-head to regulate the flow of oil and gas.

Cilia are microscopic hairlike threads projecting from the surface of a cell. They beat rhythmically enabling the cell to move through water or, if the cell is fixed, drawing water over its surface. In the BIVALVES such as the mussel, cilia on the thick fleshy GILLS help to pass food particles forward to the mouth.

Coastal currents – see LONGSHORE DRIFT

Coelenterates are a large diverse group of comparatively simple animals which have a digestive cavity with only one opening. The group includes JELLYFISHES, CORALS and HYDROIDS.

Commensalism is an association between two or more individuals of different species from which one derives feeding or other benefits without significantly affecting the other. A typical example is the wide range of animals, the most conspicuous being the sea anemone *Calliactis parasitica*, that are often associated with the hermit crab *Pagurus bernhardus* – a crab found in shallow water around the British Isles.

Continental crust is the thickened portion of the Earth's crust which forms the continental blocks. Typically it is about 35 kilometres (21 miles) thick and consists primarily of granitic rocks.

Continental drift refers to the theory that the continents have moved, and are still moving, around the Earth's surface. According to this theory, the Earth's crust is split into a number of plates which support the continents. These plates are moved about essentially as a result of CONVECTION CURRENTS within the Earth's MANTLE. New crustal material is added at the MID-OCEANIC RIDGE in the divergence zone where plates are moving apart. Elsewhere, in the convergence zone, plates collide and either become compressed leading to mountain building, or one plate slides beneath the other producing a deep OCEANIC TRENCH. Volcanoes and earthquakes are common in both divergence and convergence zones.

Continental rise – see CONTINENTAL SHELF

Continental shelf is the extension of the continental land mass into the ocean. It slopes very gently to a depth of around 180 metres (600 feet) when the gradient steepens to between 4 and 20 degrees at the start of the continental slope. This in turn leads to the more variable continental rise and, beyond, the abyssal plains of the deep ocean floor. The width of the continental shelf varies considerably; for example, it is virtually absent off the western edge of South America – a convergence zone – but extends for several hundred kilometres on the east coast of the continent off Argentina.

A large Caribbean coral with polyps out.

Divers decompress by stopping for a time at different depths during their rise to the surface.

Continental slope – see CONTINENTAL SHELF

Convection currents are the means by which heat moves from one place to another in gases or fluids, including molten rock. When a gas or liquid is heated, particles near the heat expand and become lighter (less dense). As a result they rise towards the top of the mass of gas or liquid. As they approach the surface they spread and begin to cool down. Finally, they become so dense that they begin to sink again to complete the cycle. Convection currents in the semifluid rocks underlying the Earth's crust are thought to be an important mechanism in CONTINENTAL DRIFT. They also play an important part in the circulation of ocean CURRENTS.

Convergences occur where circulating water masses come together (converge), usually resulting in a sinking of the surface water.

Copepods, although CRUSTACEANS, are sometimes referred to as 'the insects of the sea' because of their widespread distribution and great abundance. These tiny shrimplike animals constitute an important part of the marine FOOD WEB, particularly in temperate and sub-Arctic waters.

Corals are marine COELENTERATES, occurring either singly or in extensive colonies, which form a hard skeleton of calcium compounds or other materials. The corals which form large REEFS are limited to warm, shallow and clear waters, but those forming solitary, much smaller, growths may be found in colder waters to great depths.

Coriolis effect is produced by the Earth's rotation. Moving particles are deflected to the right in the northern hemisphere and to the left in the southern hemisphere. Together with the PREVAILING WINDS and the Sun, it is largely responsible for the pattern of oceanic circulation.

Crustaceans are arthropods – animals with a tough segmented external skeleton and with jointed legs and other appendages. They breathe by means of GILLS or similar structures. The group includes barnacles, crabs, shrimps and lobsters.

Currents in the oceans are basically the result of three factors – the PREVAILING WINDS, the Earth's rotation (combined with the location of the land masses), and the heating of the Earth's surface which helps to create CONVECTION CURRENTS.

D

Decca is a British navigational system which provides position by comparing continuous synchronized signals from two fixed transmitters.

Decompression is the systematic reduction of pressure during a diver's ascent. It is generally carried out by making stops according to a predetermined schedule. Decompression sickness ('the bends') occurs when the diver ascends too quickly: the rapid release of the pressure allows gases dissolved in the body fluids and tissues to form bubbles before reaching the lungs. The only satisfactory remedy for decompression sickness is to recompress the diver, otherwise bubbles formed in vital tissues will prove fatal.

Decompression sickness – see DECOMPRESSION
Deep scattering layer – see PHANTOM BOTTOM

A desalination plant provides fresh water for the hot desert country of Qatar.

Deep-sea mining is still in its infancy, but systems are being developed to collect MANGANESE NODULES from the deep ocean floor.

Potential 'mine sites' have already been identified in the Pacific. Prototype dredging systems have undergone trials at sea. The most successful seem to be air-lift and hydraulic dredges. These lift the nodules either by injecting air into the pipe connecting the dredgehead with the mining ship or by using a series of pumps set in the pipe. In a third system, two ships drag a long continuous line of dredge buckets along the ocean floor. The buckets are kept in continuous rotation from the ocean floor to the surface mining ships. A more futuristic, as yet untried, proposal is to use small mining machines that would shuttle between a large surface production platform and the deep ocean mine.

Demersal fish are bottom-living fish, as opposed to surface-living or PELAGIC FISH.

Desalination is the production of freshwater by the removal of salt and other mineral impurities.

Detritus consists of small particles of dead organic matter, plus the live microbes that are decomposing the material. It is often an important source of food for animals living on and in the seabed.

Diatoms are microscopic ALGAE which have a shell (frustule) composed of silica. Apart from being important members of the PLANKTON, deposits of their shells form vast quantities of OOZE on the deep seafloor.

Dinoflagellates are microscopic organisms which may possess features of both plants (such as chlorophyll and cellulose plates in the cell wall if present) and animals (such as the ability to ingest solid food).

85

Distillation is the traditional way of separating fresh water or salt from seawater. The heat of the Sun may be sufficient to evaporate seawater in shallow salt pans and leave the salt, but it is not usually effective where the aim is to extract large quantities of fresh water. In multistage flash distillation, the most frequently used desalination system, seawater is heated and fed into a succession of chambers at progressively lower pressures. As the boiling point of water becomes lower as the pressure is reduced, some of the hot water 'flashes' to steam when entering each chamber, leaving the salt behind. Flash distillation may be combined with power production to take advantage of preheated seawater from the generators.

Divergence is the horizontal flow of water in different directions from a common centre or zone. A typical example is the UPWELLING which occurs as a result of the interaction between wind and currents and leads to surface water moving away to be replaced by water from below.

Diving mixture – see SATURATION DIVING

Diving suit, invented in the early 19th century, enabled people to work in shallow waters. It consists of a watertight covering attached to a metal helmet fitted with transparent port-holes. Air is pumped down to the helmet through a tube. In order to stay submerged the diver must wear a heavy weight-belt and weighted boots. The suit, which allows only limited movement, has now been largely replaced by SCUBA diving equipment. Nevertheless, in recent times a series of 'armoured' diving suits have been designed and built. These suits are made of metal and glass-fibre reinforced plastics (GRP) and have jointed legs and arms fitted with manipulators. Wearing such a suit, a diver can descend to considerable depths without the need to decompress or use BREATHING MIXTURES. The suit's 'shirt-sleeve' environment is kept at the same atmospheric pressure as on the surface.

Doldrums is a nautical term describing the belt of light and variable winds near the equator.

Dynamic positioning refers to the active maintenance of STATION. It applies equally to ships and platforms. Very often dynamic positioning is controlled by a computer which operates propulsion units in order to maintain position with respect to a fixed reference point determined from a navigational satellite or a sonar or radio beacon.

Dysphotic zone extends down below the EUPHOTIC ZONE of the sea to about 1000 metres (3280 feet). It is the zone in which sunlight, though present, is insufficient for photosynthesis.

E

Ebb tide refers to the period during which the tidal current moves away from the shore or down a tidal stream.

Echinoderms have either an external shell-like skeleton consisting of calcareous plates with projecting spines, or plates and spines embedded in the skin. The adults have radially symmetrical, usually five-rayed, bodies. They generally move about the seabed by means of numerous tube-feet – small sac-like organs that can be extended when filled with water from a complicated circulatory system. Examples of the major divisions of living echinoderms are the starfish, sea urchins, crinoids and sea cucumbers.

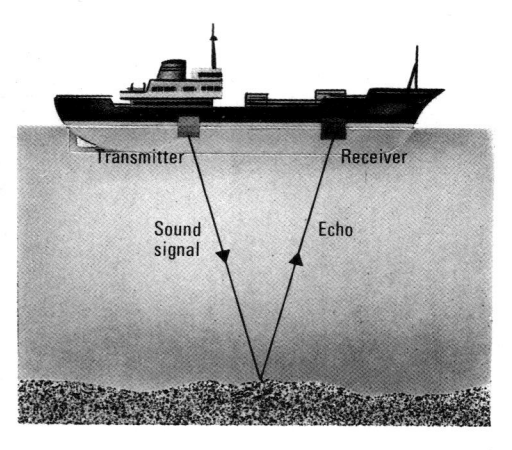

An echo sounder maps the ocean floor.

Echo-sounder is an instrument used to measure the depth of the sea. Sound waves transmitted through the water are reflected back from the seabed as echoes. By knowing the speed of sound through the water, the depth of the water can be fixed from the time taken between the emission of a sound and the reception of its echo. Most ships are fitted with some form of echo-sounder. In the 1920s an echo-sounder was used by the German research vessel *Meteor* to make one of the first systematic surveys of the ocean floor. The ocean floor proved to be far more irregular than indicated by earlier isolated SOUNDINGS.

Ecosystem is the basic unit in ECOLOGY. It includes both organisms and their non-living environment.

Electrical fishing employs electricity either to stun or to stimulate fish so that they can be caught more easily. For example, a shrimp TRAWL has been devised which uses pulses of electricity to make shrimps move from their mud burrows into the water above. Electrical fishing can also take advantage of the so-called *anodic effect*. When subjected to an electrical current fish tend to move towards the positive pole or anode.

Electrodialysis uses selectively permeable membranes to remove salt from seawater. The driving force is electricity.

Euphotic zone of the sea (also called *photic* zone) extends to a depth of about 50 metres (164 feet) in middle latitudes and about 100 metres (328 feet) in the tropics. It is the zone in which the amount of light is sufficient for photosynthesis.

Eustatic changes in the sea level are the result of changes in the capacity of the ocean basins or the volume of the oceans, caused for example by fluctuations in the volume of glacier ice, sedimentation and earth movements.

Eutrophication results from the 'overfertilization' of river, lake and sea. In nature, the process occurs slowly as shallow lakes fill with sediment, but POLLUTION accelerates it, usually with undesirable results. Typically, the release of excessive quantities of nutrients such as nitrates and phosphates stimulates algal BLOOMS. The subsequent decomposition of these plants depletes the water of oxygen. A thick decaying sludge builds up, suffocating all but the hardiest creatures.

Exclusive Economic Zone (EEZ) is the area offshore, the resources of which fall under the jurisdiction of the coastal state.

F

Fathom is a common unit of depth (equivalent to 1·83 metres or 6 feet).

Fetch refers to a continuous area of water over which the wind blows uninterruptedly in the generation of the ocean SWELL.

Filter-feeding is common among marine animals. Instead of actively pursuing their food, they filter or trap edible particles from the surrounding seawater.

Fjord is a narrow, deep, steep-walled inlet of the ocean formed either by the flooding of a mountainous coast or by the ocean entering a deeply excavated glacial trough after the glacier has melted away.

Floe is sea ice in either a single unbroken piece or individual pieces spread over an area of water.

This spiny starfish is an example of an echinoderm.

Food chain – see FOOD WEB
Food pyramid – see FOOD WEB
Food web describes the pathway of energy often referred to as a *food chain*, from primary producers which capture sunlight through the process of photosynthesis to the secondary TROPHIC LEVEL, the herbivores, and on to the tertiary and other levels, the carnivores. Each linking point in the mesh of the web is occupied by one or a group of species which have in common the way they feed and the way they are preyed upon by animals at a higher level. At each of the points in this *food pyramid*, energy 'changes hands'. In the process as much as 90 per cent of it is lost.

Freezing processes for desalination exploit the fact that when seawater freezes, freshwater ice forms, leaving the salt in solution. Basically there are two processes: direct freezing where the water is frozen by evaporation in a vacuum chamber; and secondary refrigerant freezing which uses the evaporation of a liquefied gas to cool the seawater to the freezing point.

Fringing reef is a REEF attached to an insular or continental shore.

G

Gametes are the male and female reproductive cells. Male gametes are usually small and motile (sperm); female gametes (eggs) are larger and contain yolk.

Gametophyte – see ALTERNATION OF GENERATIONS

Geneva Continental Shelf Convention (1958) provides a legal framework within which the continental shelves of the world can be explored and exploited. It defines the CONTINENTAL SHELF as the seabed and subsoil of the submarine areas adjacent to the coast but outside the area of the territorial sea to a depth of 200 metres (656 feet) or, beyond that limit, where exploitation of the natural resources is possible. This 'open-ended' definition has been criticized, particularly by developing countries, as giving countries with advanced technology a free hand to exploit the oceans, even the deep ocean floor. The Convention came into force in 1964 after the 22nd nation, the United Kingdom, signed it.

Geophysical exploration employs gravitational, magnetic and seismic techniques. Gravitational methods are based on the way in which the Earth's gravitational pull on an object suspended above it varies from place to place according to the density of the underlying material and its distance away from the object. Magnetic techniques measure the changes in the Earth's local magnetic field induced by the magnetic minerals that rocks contain. Seismic methods give the fullest picture of the seabed. In seismic surveying an artificial 'earthquake' is created, sending out waves of energy which surge downwards through the seabed. The velocity at which these shock waves pass through the rock depends on its elasticity, density and composition. Whenever these properties change – for example, from one rock layer to another – some of the energy is reflected back to the surface. These 'echoes' are picked up by 'geophones' to give a picture of the various layers and structures within the sea-floor.

One of Norway's many fjords, carved out by glaciers and flooded by the sea.

Gill net is a fishing net that traps PELAGIC FISH when they endeavour to swim through its meshes. Fish which are too large to pass through the mesh become entangled by their gill covers when they try to back out of the net.

Gills are delicate, thin-walled structures used for the exchange of respiratory gases in a water environment. Very often, the surface of the gill is folded to increase its area and improve the gas exchange. The efficiency of the gill may also be increased, as for example in fishes, by having the blood inside the gills flow in an opposite direction to the water on the outside.

Gravel consists of loose fragments of material ranging in size from approximately 2 to 256 millimetres (up to $\frac{1}{16}$ inch).

Green algae (Chlorophyta) derive their colour from the chlorophyll that they contain. They occur predominantly in fresh water, but there are a few marine specimens, for example the sea lettuce *Ulva*.

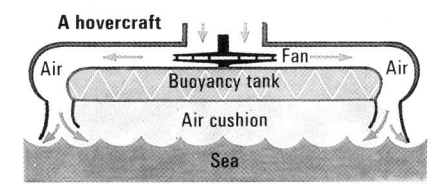

A hovercraft

Ground-effect principle is used by the hovercraft and similar machines. If a stream of air is directed through a hole in a flat metal plate down onto a surface, it creates an upward thrust which lifts the plate.

Guyot is a flat-topped submarine mountain.

Gyres are the loops, each one almost the full width of the ocean, traced by the major surface CURRENTS. In the northern hemisphere there are two gyres – one in the Atlantic and the other in the Pacific – rotating slowly in a clockwise direction. South of the equator there are three counter-clockwise gyres – in the Atlantic, Indian and Pacific Oceans – and a fourth one which surrounds the Antarctic continent.

H

High pressure nervous syndrome (HPNS) describes a group of disorders which appear when divers are compressed to high pressures. The onset of the symptoms, which include muscular tremors and spasms, sleepiness, visual disturbances, dizziness and nausea, seems to depend upon the speed of compression and the pressure to which the person is subjected. Symptoms appear progressively, but they do not usually affect mental performance although the involuntary tremors can hamper manual work. The effects can be overcome, but the HPNS may help set the limit to which divers can safely descend in the sea.

Hovercraft – see GROUND-EFFECT PRINCIPLE

Hydrocarbons are organic chemical compounds composed only of carbon and hydrogen. They are the principal constituents of petroleum and natural gas. They provide vital raw materials for the production of plastics, solvents and industrial chemicals.

Hydrofoil craft travel on wing-like projections immersed in the water which reduce the area of boat in contact with it. The reduced drag enables hydrofoil craft to attain higher speeds than conventional displacement vessels.

Hydrofoils are used for fast travel over short distances.

Inertial navigation instruments help ships to navigate without external reference points, such as stars or magnetic north.

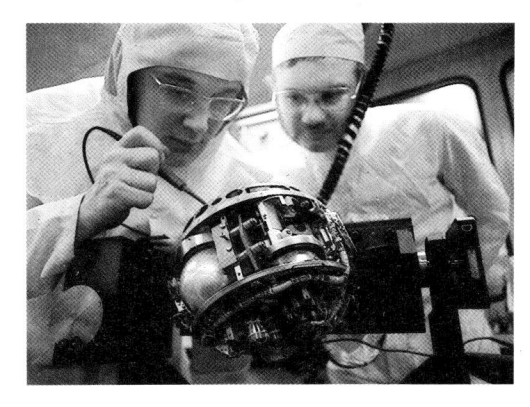

Hydrography is concerned with the measurement and mapping of the physical features of the oceans.

Hydroid is the POLYP form of COELENTERATE. It is a sessile (attached) animal which produces pelagic JELLYFISH by asexual budding. (See ALTERNATION OF GENERATIONS.)

Hydrophone picks up water-borne sound waves.

Hydrosphere includes all the water on the Earth, of which about 97·21 per cent is in the oceans; 2·15 per cent is ice; 0·619 per cent is ground water; 0·02 per cent is in rivers and lakes; and 0·001 per cent is water vapour in the atmosphere.

I

Iceberg is a mass of ice that has broken away from land and floats in the sea, or becomes stranded in shallow water.

ICES (International Council for the Exploration of the Sea), one of the earliest international organizations concerned with marine research, was established in 1902 to assist the fisheries of western Europe. It still provides a focal point for marine biological research and the collection of fishery data. It advises national and international fishery organizations on the state of commercial fish stocks. Above all, it provides a neutral forum for the fishery scientists of member nations.

Ice shelf is a thick ice formation with a fairly level surface, formed along a polar coast and in shallow bays and inlets. It may extend hundreds of kilometres out to sea.

IMCO (Intergovernmental Maritime Consultative Organization) is a specialized agency of the United Nations whose activities are entirely in the maritime field. Established by a convention in 1948, it came into being in January 1959. The Organization's objectives are to assist cooperation among governments in technical matters of all kinds affecting shipping engaged in international trade and to promote high standards of maritime safety and navigation. IMCO's headquarters are in London.

Indicator species are members of the PLANKTON that can be used to identify particular masses of water. Because many of the species are widespread, identification of the water usually depends on the association of the various species rather than on individual ones.

Inertial navigation system, once set, gives position without any outside reference. It operates by detecting changes in direction and speed by continuous reference to accelerometers mounted on a stable platform provided by three gyros.

International Ice Patrol was established as a direct result of the sinking of the ocean liner *Titanic* on 11th April, 1912, when 1513 people died. The Patrol, which is based in Newfoundland, watches and reports on ice movements as well as other obstacles in shipping lanes that might prove hazardous in fog. The ice generally becomes a serious menace in March when the Arctic floes begin to break free. The danger rarely persists beyond June or July.

Intertidal (littoral) zone is generally defined as the zone between mean high-water and mean low-water levels.

IOC (Intergovernmental Oceanographic Commission) was established in 1961 within the framework of Unesco (United Nations Educational, Scientific and Cultural Organization) to promote OCEANOGRAPHY at governmental level, with special regard to the developing countries. The aim of IOC is to secure progress in oceanic investigations for the peaceful use of the oceans and their resources by the concerted action of member states.

Isobath is a depth contour.

Isostasy refers to the equilibrium maintained by portions of the Earth's crust floating on the denser MANTLE below. The isostatic theory implies that large continental masses are supported by 'roots' of low-density crustal material.

J

Jack-up platform is a platform supported by legs resting on the seabed. The platform can be floated into position and then raised above the surface by lowering the legs to the seabed. Jack-up platforms are ideal for drilling in shallower waters. They also provide work and engineering bases for offshore construction.

Jellyfish is a free-swimming form of COELENTERATE. It has a disc- or bell-shaped body of a jellylike consistency. Many, such as the Portuguese man-of-war *Physalia*, have long tentacles with stinging cells.

K

Kelp is a common name for the brown seaweed found fringing most rocky shores. It belongs to a group, the laminariales, which includes the largest known seaweeds. See also BROWN ALGAE.

Knot is a unit of speed used in navigation. It is equivalent to one NAUTICAL MILE (1852 metres or 1·1508 land miles) per hour.

Euphausiid, a member of the krill family.

Krill are shrimplike, planktonic CRUSTACEANS, and the major source of food for baleen WHALES. Krill abound in the waters of the Southern Ocean and are being exploited increasingly for conversion into meal for animal feeds as well as products suitable for human consumption. The total annual yield could eventually be as high as 60 million tonnes per year. This is only a little less than the total annual world catch of marine fish during the 1970s.

L

Lantern fishes (myctophids) are small oceanic fishes which normally live at depths between a few hundred and a few thousand metres. They are one of the few major fish resources that remain unexploited.

Lateen sail was first used in the eastern Mediterranean about the 2nd century AD. It consisted of a triangular sail slung from a long sloping yard (crossbar) attached to the top of the mast so that its peak was much higher than the mast but the throat of the sail was almost level with the deck. This rigging allowed the sail to take wind on either side enabling the vessel to tack into the wind. A combination of the lateen with the existing square sail led to the ocean-conquering sailing ship, the CARRACK, of medieval times.

Lay barge is a large ocean-going vessel used to lay oil and gas pipelines on the sea-floor. Generally the barge carries a cargo of lengths of pipe which are welded together and fed out from the stern as it moves forward. If the sea-floor is deeper than about 60 metres (197 feet), the pipeline is supported by a submerged 'stinger' which is trailed behind the barge.

Light penetration in the sea varies according to wavelength and the 'cloudiness' of the water. All productivity in the sea depends ultimately on the availability of light for photosynthesis, starting point of the FOOD WEB. A great deal of light is reflected at the surface especially when it is ruffled. Even in calm clear water, 3 to 4 per cent of light may be reflected at the surface. Of the remainder, up to 80 per cent is absorbed in the top 10 metres (33 feet) or so. In clear water the blue-green part of the spectrum travels furthest.

Long-line fishing uses a long weighted line with side branches ending in hooks to catch mainly DEMERSAL and midwater fish.

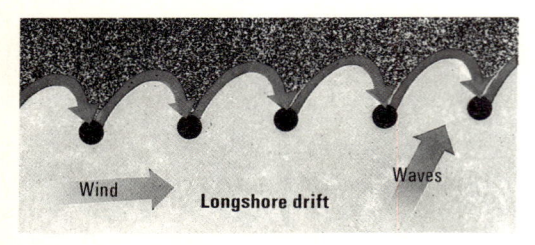

Longshore drift refers to the transport of material along the shore by a current produced by waves being deflected at an angle to the shore. Material carried by this longshore current is usually deposited at points where obstructions cause it to lose speed and hence carrying power.

Loran (Long Range Navigation) is an American navigational system which fixes ship position by reference to the difference in time of synchronized radio signals from two fixed transmitters.

M

Magellan, the Portuguese sailor, who in 1519 led the first expedition to sail round the world. He was killed on the way.

Manganese nodules are considered to be the ore of the deep. They consist mainly of manganese and iron, but also contain significant quantities of copper, nickel, cobalt and traces of other elements. The minerals contained in the nodules all occur in seawater, but at much lower concentrations. Cobalt, nickel and copper are enriched a millionfold or more over the surrounding water. The very rare radioactive element thorium is concentrated 100-millionfold.

The nodules are widely distributed in the world's oceans, but the deposits most favoured for DEEP-SEA MINING are found in the Pacific Ocean, just north of the equator, at a depth of about 5000 metres (16,400 feet). The total quantities of metals held in nodules is immense and, unlike minerals on land, is being added to continuously. Every year, for example, some 55,000 tonnes of copper are added to nodules on the floor of the Pacific Ocean alone. The individual nodules grow very slowly, however, adding between 1 and 100 millimetres per million years – equivalent to about a layer of atoms per year.

Mantle is the part of the Earth lying between the crust and the core. It is 2900 kilometres (1800 miles) thick. Near the top, some of the rocks are semi-fluid and move in CONVECTION CURRENTS.

Marigram plots the rise and fall of the tide against time.

Maximum sustainable yield is the maximum catch that a fishery can sustain for many years without depleting the stock.

Mid-oceanic ridges rise like mountain ranges from the ocean floor forming islands in places where they break the surface of the sea. A vast chain of mid-oceanic ridges over 60,000 kilometres (37,000 miles) long extends all around the globe, through every ocean. The mid-oceanic ridge generally rises about 3000 metres (10,000 feet) above the abyssal plain. A central valley runs along the top. This is the site where new material is added to the sea-floor as a result of CONVECTION CURRENTS in the Earth's MANTLE – the basic cause of CONTINENTAL DRIFT.

Mineral-rich sands have become a primary target for marine mining. They occur under shallow water on the CONTINENTAL SHELF, are within the range of conventional dredging techniques, and have the potential to supply a number of vital minerals. Minerals exploited include: heavy metals – ilmenite and rutile (sources of titanium), zircon (for production of the zirconium widely used in foundry sands and refractories), monazite (rare earth phosphate mineral containing up to 12 per cent thorium), chromite (principal ore of chromium), magnetite and iron, and tin; sand and gravel, aragonite (a particularly pure form of calcium carbonate), shells, phosphorite and coal; and precious metals and minerals – gold, diamonds and precious coral. Australia is the country with the most extensive beach mining and dredging operations.

Molluscs are among the most conspicuous invertebrate animals and include such familiar forms as mussels, oysters, squids, octopods and snails. Typically they have a body protected by a single or two-hinged shell, but some molluscs such as the sea slugs and octopus have lost the shell completely. The molluscs comprise the largest group of invertebrates after the arthropods (insects, crustaceans, spiders etc). Over 80,000 living species have been described. In addition some 35,000 fossil species are known.

Monsoons are winds which reverse direction from season to season. The term, which is derived from the Arabic *mausim* (season), was first applied to the winds over the Arabian Sea, which blow from the north-east for six months and then from the south-west for the remainder of the year. It is now used to describe similar wind systems elsewhere.

N

Nannoplankton is PLANKTON with a diameter of less than 50 microns. It is so small that individuals pass through most plankton nets. Nannoplankton is usually collected by centrifuging samples of seawater.

Nautical mile is the length of one minute of arc along any great circle on the surface of the Earth. Since this varies with latitude, the nautical mile is agreed internationally to be equal to 1852 metres (6076·103 feet or 1·1508 statute miles).

Navigational instruments enable the navigator to determine his position without reference to the land and to chart and follow courses across the open ocean. Prior to their invention and development, navigators rarely strayed deliberately far from the shore. Initially when out of sight of land, navigators were guided by the Sun and stars, but from the 11th century a series of inventions began to transform navigation. Early among these was the introduction of the magnetic compass. The sand-glass combined with the log, a float on a line of known length released from a moving ship, enabled the measurement of speed and hence distance. Modern navigational instruments include electronic logs, and INERTIAL NAVIGATION SYSTEMS that, once set, continue to indicate a ship's position without further outside reference. Ships may also be able to fix their position from navigational satellites orbiting the Earth, or by navigational systems such as DECCA and LORAN.

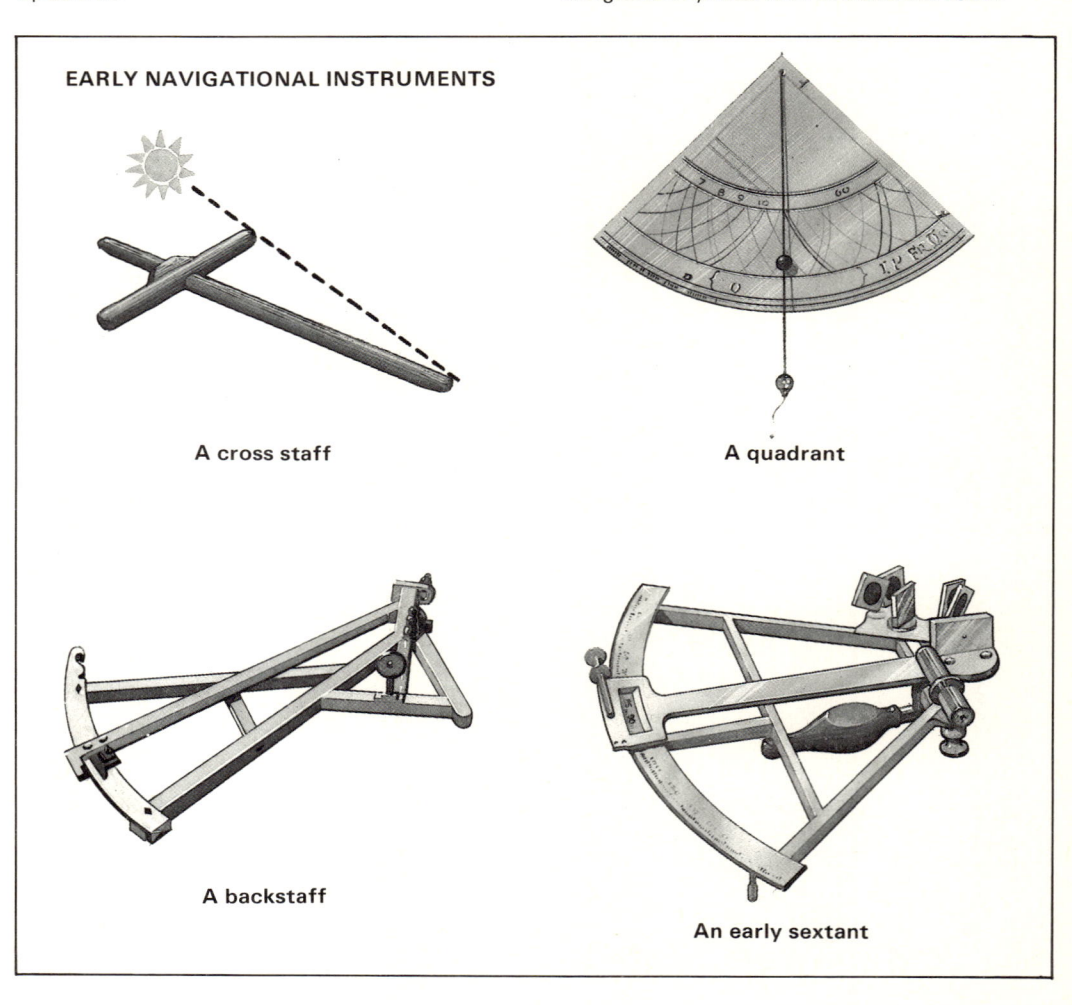

EARLY NAVIGATIONAL INSTRUMENTS

A cross staff

A quadrant

A backstaff

An early sextant

Neap tide – see TIDES

Nekton is the collective name for all animals which swim in the sea.

Nitrogen narcosis describes the effect on a diver's mental processes of excess nitrogen dissolved into his body fluids and tissues. When a SCUBA diver breathing normal air, which contains 80 per cent nitrogen, descends much below 18 metres (60 feet), the nitrogen begins to impair his performance. The narcotic effects become progressively worse with increasing depth so that by about 90 metres (300 feet) even routine tasks become practically impossible. For deep dives, the nitrogen is largely or completely replaced by an inert gas, usually helium. The amount of oxygen in the breathing mixture is also reduced since this gas, normally considered a life-giver, becomes toxic under pressure if present in the proportions found in air.

O

Ocean basin is that part of the floor of the ocean that is more than about 2000 metres (6500 feet) below sea level.

Oceanic trench marks a *subduction zone* where two of the plates that constitute the Earth's surface (see CONTINENTAL DRIFT) collide and one dips down into the MANTLE. They are the deepest parts of the ocean. Evidence for the forces at work are the earthquakes and volcanoes that occur in the Benioff Zone. This zone is named after the American geophysicist who first described the way in which the edge of an oceanic plate dips down at an angle of about 30 degrees until at a depth of about 700 kilometres (435 miles) it breaks up and finally melts within the inner mantle of the Earth.

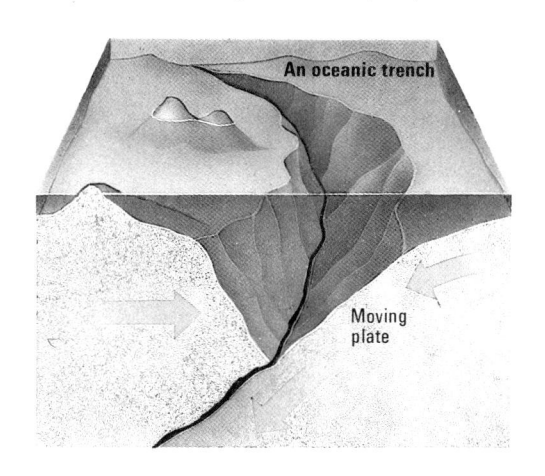

Oceanography is the application of all sciences to the study of the sea. The British Challenger Expedition (1872–76) is generally considered to mark the beginning of modern oceanography.

Ooze is a fine-grained deep-ocean sediment composed largely of the shells and undissolved remains of foraminifera, DIATOMS and other marine life.

Over-fishing – see MAXIMUM SUSTAINABLE YIELD

Oceanographic research is carried out by this strange ship called FLIP, the Floating Instrument Platform. It is equipped with ballast tanks that can be flooded to sink the stern when it reaches its site.

Divisions of the Sea

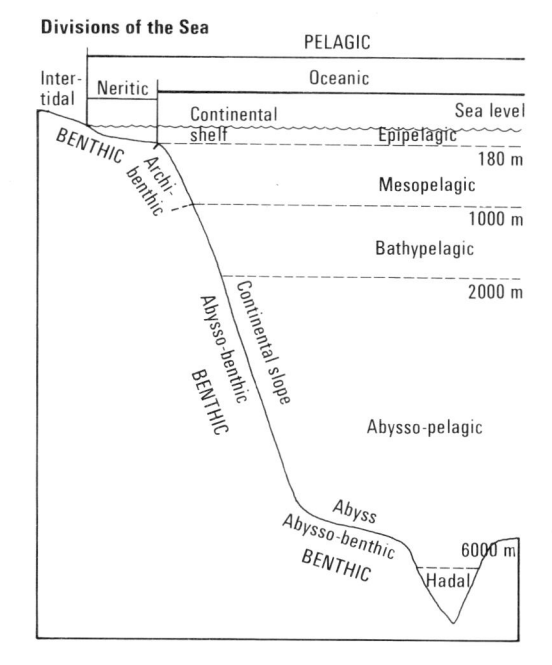

P

Pack ice refers to any large area of ice FLOES that have been driven together to form a solid mass, thereby preventing or obstructing navigation.

Pangaea is the name of the single 'super continent' thought to have existed 200 million years ago. About 180 million years ago, it started to break up and the continents began drifting over the Earth's surface, a process that is still continuing. See also CONTINENTAL DRIFT.

Pelagic division is a primary division of the sea which includes the whole mass of water.

Pelagic fish are found in the surface waters of the sea, as opposed to bottom-dwelling or DEMERSAL FISH.

Pelagic sediments are found on the ocean floor remote from any land. They are generally very fine and range in colour from white to a dark reddish-brown. Pelagic sediments containing less than about 30 per cent of organic remains are called red clay, while those containing more than this are known as OOZES.

Phantom bottom is a false bottom indicated by SONAR. Unlike recordings of echoes from the sea-floor, the echoes from this layer are diffuse and ill-defined. When discovered in the early 1940s, the source of this sonar disturbance was called the *deep scattering layer*. Subsequent investigations have revealed the presence of several layers which vary in depth according to the diurnal migrations of marine animals.

Photic zone – see EUPHOTIC ZONE

Phytoplankton – see PLANKTON

Plankton is the collective name for the floating, drifting life found in the sea. It is subdivided into the *phytoplankton* – the plants, and the *zooplankton* – the animals. Ultimately all life in the sea relies on the ability of the phytoplankton to carry out photosynthesis. Because this involves the use of solar energy, the microscopic plants such as DIATOMS are restricted to the sunlit zone of the surface waters. The zooplankton, however, can be found at most depths. Frequently members of the zooplankton migrate upwards during the evening to feed on the phytoplankton, retreating deeper with the dawn.

Pollution in the oceans results solely from people and their activities. Never before have chemical pollutants entered the marine environment in quantities sufficient to affect the natural chemistry of the oceans. People have increased the quantity of naturally occurring chemicals reaching the oceans and have added synthetic products which have no natural counterparts. They have also interfered with the physical state of coastal waters by releasing warm water effluent from power stations and factories, so causing thermal pollution, or by replacing natural marine habitats with concrete quays and holiday resort developments. In addition, the natural fauna and flora have been changed by the release of untreated sewage, the deliberate introduction of commercial species into new areas and the accidental introduction of pest species from elsewhere. Pollution can, therefore, be chemical, physical or biological in nature.

Polychaetes are segmented marine worms, some of which are sedentary and live in tubes, while others are active swimmers and burrowers. Polychaete worms are very common marine animals, but they usually remain unnoticed because of their secretive habits. Among the most beautiful of the sedentary polychaetes are the fanworms such as *Sabella pavonina*, the peacock worm. An interesting free-moving polychaete is the sea mouse *Aphrodite aculeata*, the back of which is covered with what looks like iridescent fur.

Polyp is a sessile (attached) form of COELENTERATE. Typically, the body consists of a cylindrical stalk attached at one end to an object and with a mouth surrounded by tentacles at the other. The sea anemones are examples of polyps. They are usually only a few centimetres long, and perhaps 3 centimetres (just over an inch) in diameter, but specimens of *Stoichactis* on the Great Barrier Reef of Australia may have a diameter of a metre (3·2 feet) or more. Sea anemones are often brightly coloured. This and the numerous tentacles can lead the casual observer to regard them as flowers rather than animals.

Prevailing winds are winds that blow from the same direction for most of the time. They are a product of planetary rather than local forces. They are the north-east and south-east TRADE WINDS, the westerlies, and the polar easterlies. The prevailing winds play an important part in generating ocean CURRENTS.

Protozoans are mostly microscopic, one-celled animals. The protozoan, although consisting of a single cell, is equivalent to a multi-cellular organism. It is a complete organism which carries out all the functions found in any larger animal. Protozoans constitute one of the largest populations in the oceans.

R

Radiolarians are planktonic PROTOZOA possessing a skeleton of silica. They may be an important constituent of deep ocean OOZES.

Red algae (Rhodophyta) derive their colour from a predominance of red pigment over the other pigments present. Almost all the species are marine. The plants rarely exceed several centimetres in length. They show an ALTERNATION OF GENERATIONS with a well-developed gametophyte. Many of the common red algae of the seashore have a life history unknown among other algae, or any other plants. Following fertilization, the zygote develops into a multicellular plant attached to the gametophyte. It then produces spores which grow into plants resembling the gametophyte generation. These in turn release spores to produce the dominant gametophyte plants. There are thus three generations in all.

Red clay – see PELAGIC SEDIMENTS

Red tide describes a bloom of DINOFLAGELLATES (members of the PHYTOPLANKTON) in which the plants become so numerous that the surface water becomes coloured red. Where the predominant dinoflagellates are highly toxic, for example *Gonyaulax*, the ensuing red tide may kill many sea birds and fish.

Reef is a rock, often consisting of coral, which lies 20 metres (65 feet) or less from the surface of the sea and presents a hazard to navigation.

Refraction of water waves is the process by which waves moving into shallow coastal water come to conform with the contours of the sea-floor. Waves moving in the shallower parts of the sea-floor are slowed down compared with those in the deeper water causing the wave 'front' to slew round.

A seabird with its feathers covered in oil – a victim of pollution from the *Bohlen* tanker disaster of 1976.

Reproduction of marine animals and plants show the same range and variety as on land with the important exception that few flowering plants live in the marine environment. The dominant form of plant life (ALGAE) does not have flowers but shows a similar ALTERNATION OF GENERATIONS in the life cycle. Most marine animals reproduce sexually, but some species are able to increase their numbers by asexual budding. They may also show an alternation of generations more marked than that normally associated with land animals.

Reverse osmosis produces fresh water by 'squeezing' seawater under pressure through a membrane which is impermeable to salt and similar impurities.

Ria is a narrow inlet of the sea. Rias are flooded river estuaries produced either by a rise in sea level or sinking of the land.

Rotary drilling is the system generally used to penetrate underground oil and natural gas reservoirs. In this method, a drill bit rotates while being pressed down on the bottom of the borehole. The rock is gouged and chipped away. The bit is connected to the surface equipment through a drill pipe surrounded by a sleevelike casing. Drilling mud is passed down through the pipe and returns to the surface inside the casing bringing with it the rock chippings. The drilling mud also prevents fluids from the formation entering the pipe to cause a 'blow-out' or 'gusher'. The drill pipe is usually rotated from the surface. It has a square or octagonal section at the top called the kelly which keys into a hole of a similar shape in a turntable. The circulating drilling mud, which is filtered of rock debris at the surface, enables continuous drilling, but the drill bit wears rapidly and needs frequent replacement. The drill pipe must then be raised from the borehole which is sealed with the heavy drilling mud.

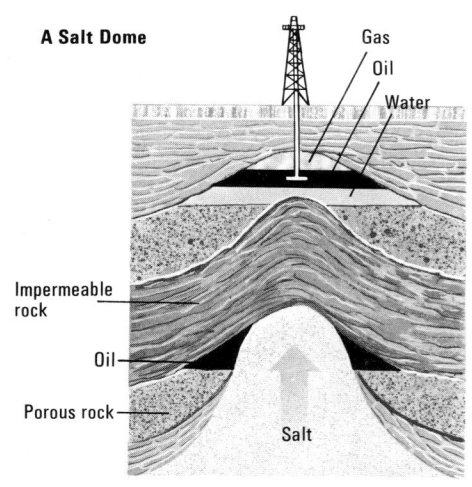

S

Saline water has a salt content of 17,000 parts per million or more.

A Salt Dome

Salt dome is a mass of underground salt which pushes through other layers of rock. These structures are important because HYDROCARBONS are often trapped around the top of the dome.

Saturation diving is the technique whereby a diver remains at or near the depth or the equivalent pressure at which the body tissues become saturated with the inert gas of the *diving mixture*. Since decompression time is dictated by the degree of saturation, once this point is reached, decompression time remains the same, irrespective of duration at this pressure. The method, therefore, greatly increases the ratio of time spent working to that wasted during decompression.

SCOR (Scientific Committee on Oceanographic Research) was established in July 1957 by the International Council of Scientific Unions (ICSU), to further international scientific activity in all branches of oceanic research. The first major scientific project of SCOR was the International Indian Ocean Expedition – one of the most comprehensive international research efforts ever attempted. Among its present-day responsibilities is to advise the Intergovernmental Oceanographic Commission (IOC).

Scuba stands for self-contained underwater breathing apparatus, of which there are three basic designs: the closed circuit system – a circulating, inert gas receives fresh oxygen to replace that used; the semi-closed system – a pre-mixed gas is metered out into the breathing circuit so that some is rejected continuously; and the open system – compressed air or gas is supplied from cylinders to a demand regulator and each breath is exhaled to the surrounding water.

SDC, DDC are abbreviations for the submersible decompression chamber and the deck decompression chamber that together form the chief components of modern diving systems. Divers descend into the sea in the SDC which provides a refuge and base. Usually each diver has only an emergency supply of BREATHING MIXTURE: both breathing gases and suit heating are provided by an umbilical cable from the SDC. Divers return to the surface, still at pressure, in the SDC which is attached directly to the DDC, a much larger chamber. Here divers can relax or begin the slow process of decompression to atmospheric pressure in comfort and safety.

Sea ice refers to all ice formed by the freezing of seawater. It may be called fast ice if it is attached in any way to the shore or to the bottom.

Seaweed – see BROWN ALGAE, KELP

Sediments in the ocean are subdivided into two main groups – PELAGIC and TERRIGENOUS SEDIMENTS. (See also BIOGENOUS SEDIMENT.)

Seismic sea wave – see TSUNAMI

Semi-submersible platform consists of a platform standing on submerged floats. The platform may move under its own power or be towed to its offshore drilling or work site. It may stay in position using thrusters which automatically switch on and off to maintain STATION with reference to a fixed point, usually a sonar beacon set on the sea-floor, or it may be held by anchored lines. Generally speaking, semi-submersible platforms are used in waters that are too deep for JACK-UP PLATFORMS or where the sea-floor is unstable.

Shadow zone is an area of the sea 'unlit' by conventional SONAR, caused by the bending or refraction of sound waves during their passage through water.

Sonar stands for Sound Navigation And Ranging. Originally it referred to systems using sound to determine the presence, location, or nature of underwater objects, but sonar systems are now used much more widely.

Sonobuoy is a remote 'listening post' capable of detecting underwater sounds and relaying them via radio to remote observers. Operated from aircraft, it is one of the most important means of ocean surveillance and the location of anti-submarine warfare (ASW) targets.

Sounding is the measurement of the depth of water beneath a ship.

Sporophyte – see ALTERNATION OF GENERATIONS

Spring tide – see TIDES

Station, in OCEANOGRAPHY, is the geographic location at which any set of observations are taken.

Storm surge is exceptionally high water produced by severe storms at sea. Several meteorological factors may be involved but by far the most important ones are the low pressure areas in the vicinity of the storm centre and the driving force of the high winds which pile up the water. The profile of the sea-floor or the coincidence of spring tides (see TIDES) can enhance the effect.

The *Patrick Henry* **was the second ballistic missile submarine acquired by the US navy.**

Strait is a relatively narrow waterway between two larger bodies of water. Straits have a significance in international law of the sea because as coastal states extend their controls over adjacent waters important shipping routes fall under their jurisdiction. The question of free movement through these waters is a vital one to trading nations and the maritime powers.

Subduction zone – see OCEANIC TRENCH

Submarine is an underwater vessel in which both equipment and crew are contained within a pressure hull kept at atmospheric pressure. In 1958, the nuclear submarine *Nautilus* of the United States Navy made the first submerged crossing under the North Pole. In 1960, the US Navy's *Triton* circumnavigated the world. It covered 41,500 miles in 84 days. The modern nuclear submarine armed with missiles capable of hitting enemy targets 7800 kilometres (over 4800 miles) away is an important weapon in warfare. It can deliver a 'second strike' even if its home territory has been rendered helpless by the first enemy attack.

Submersible is an underwater vessel which carries the crew in a pressure hull kept at atmospheric pressure, but stores much of its equipment outside this chamber. It usually has an overall density slightly less than that of seawater and submerges by flooding ballast tanks or, more commonly, by taking on disposable ballast. Most submersibles are small and have only limited power and underwater endurance. Few of them carry more than three people. Some submersibles carry no crew at all, but are operated from the surface. These unmanned submersibles are usually connected to the surface by a cable that supplies power as well as the control signals. Television cameras are the 'eyes' of these submersibles. Unmanned submersibles range in size from less than a metre in diameter to machines such as *Consub 2* which are over 3 metres (10 feet) long and can operate at depths down to 600 metres (2000 feet).

Subsea completion involves installing oil and gas production facilities on the seabed instead of using a surface platform. Individual wellheads are contained in underwater chambers. The chambers have a one-atmosphere environment and can be reached by a SUBMERSIBLE or diving bell which locks onto them allowing access to their equipment. The underwater wellheads may then be connected via a manifold to a single collection point which in turn may feed the oil and/or gas into a pipeline to a shore installation, a loading buoy or platform at the surface, or a production platform. The technique is still comparatively new, but in its final version all the production facilities, including those necessary for the preliminary processing, might be based on the seabed. Systems of this type represent the most likely solution for future oil production from water depths greater than 600 metres (2000 feet).

Swash is the rush of water up onto the beach following the breaking of a wave.

Swell is a long, relatively symmetrical wave produced by winds and storms occurring in distant parts of the oceans.

Symbiosis is a relationship between two species in which one or both members benefit and neither is harmed.

Jacques Cousteau in *Conshelf 1*, the first habitat used for a long stay under water. In 1962 it was tethered 10·5 metres below the surface of the Mediterranean.

T

Terrigenous deposits are composed predominantly of material from the land. These deposits cover practically all of the CONTINENTAL SHELF region.
Thallus is the fleshy body of ALGAE which is not divided as in the higher plants into root, stem and leaves. Some algae have a thallus that is sub-divided into what superficially look like leaves, others have a claw-like holdfast with which they are attached to the seabed; but none of them has an internal structure as complex as the higher plants.
Thermocline is the region in the sea where the temperature differs markedly from water either above or below it. The existence of a thermocline, either seasonal or permanent, has important implications for biological productivity as well as the use of SONAR.
Tidal wave – see TSUNAMI
Tides are the product of the astronomical forces of the Sun and Moon acting on the rotating Earth. When the two heavenly bodies pull together or are diametrically opposed the tidal currents show their greatest range – the *spring tides*. When their forces act at right angles to one another the tidal current has its shortest range – the *neap tides*. Thus, in every lunar month there are two spring tides and two neap tides with gradations between them according to the phase of the moon.
Trade winds occupy most of the tropics. They blow from the subtropical regions towards the equator, north-easterly in the northern hemisphere and south-easterly in the southern hemisphere. Their name is derived from the use made by early sailing ships of these PREVAILING WINDS.
Trawl is a fishing net pulled along through the water. One of the first trawls was the beam trawl, a cumbersome piece of gear which had a heavy cross piece to keep the mouth open. The most important development came with the use of otter boards, kite-like structures placed on either side of the trawl, which sheared away from each other to keep the mouth of the net wide open. Both beam and otter trawl were designed for catching demersal fish: they are dragged over the sea-floor. More recently trawls have been designed for fishing in the middle waters of the ocean.
Trophic level is a division in the FOOD WEB and is defined from other levels by the method of obtaining food: primary producer, primary consumer, secondary consumer, tertiary consumer. A typical chain would be phytoplankton, zooplankton, fish, man.
Tsunami consists of ocean waves produced by a seismic disturbance in the ocean floor. The waves are not usually conspicuous out over the deep ocean but they reveal their tremendous speed and energy by rising tens of metres into the air when they reach the shallows of the shore. Because tsunamis can cause considerable damage, an extensive advance warning system has been devised in the Pacific areas which suffer most from them.
Tunicates are globular or cylindrical, often sac-like animals, many of which are covered by a tough flexible coat. Some are sessile, others are pelagic. The most interesting point about these creatures is that they belong to the same broad group as man: the animals with backbones (vertebrates). The clue to this relationship rests not in the adult, but in the larvae which look very much like tiny tadpoles.
Turbidity currents are flows of suspended material down a gradient. Turbidity currents are frequently caused by the build-up of SEDIMENT on the edge of the CONTINENTAL SLOPE.

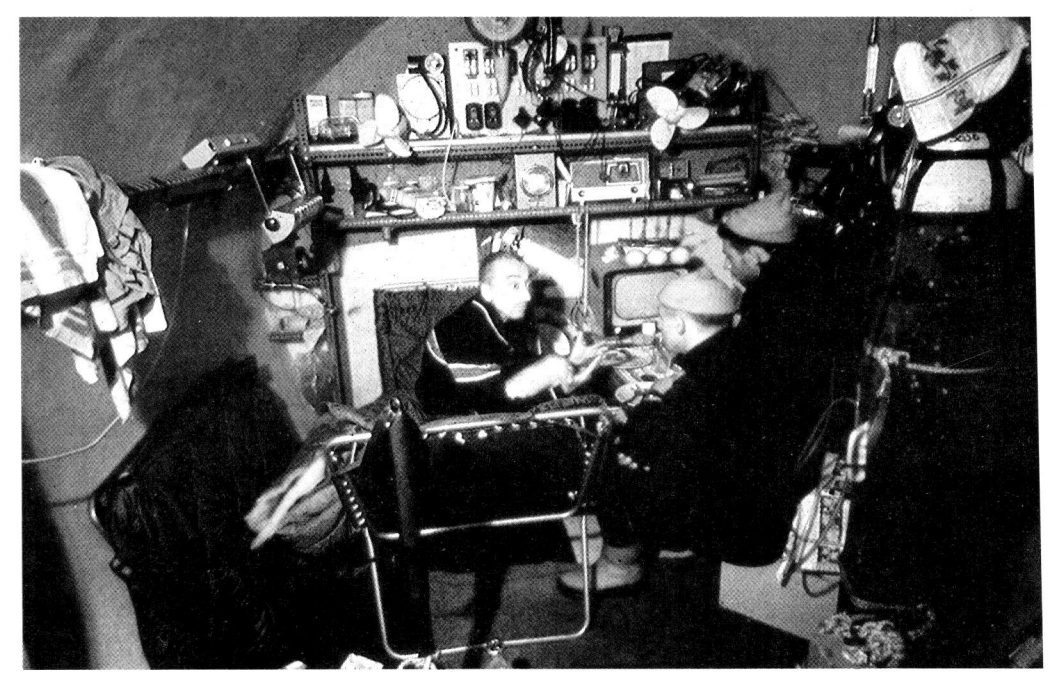

U

Underwater habitats are bases used by divers to live under the sea. Most habitats provide a dry refuge and a base from which to carry out marine research. After using all but the shallowest of these underwater homes, divers must usually undergo DECOMPRESSION in order to return to the surface.
Upwelling occurs where water rises from deep parts of the sea to the surface. Although underwater banks and shoals can deflect deep water towards the surface, the most prominent upwellings occur where the wind blows offshore or ashore. The currents generated by the wind cause the surface water to move offshore, allowing colder water to rise from below. Since deep cold water is generally rich in mineral nutrients, upwellings are usually the sites of important fisheries.

V

Vapour compression produces fresh water by compressing vapour driven off salt water using a fan. The compression and condensation of the vapour provide the heat for further vaporization of seawater in surrounding heat-exchange jackets.
Volcanic centre ('hot spot') is an area of continuing volcanic activity resulting from long-term activity within the Earth's MANTLE. These 'hot spots' remain active even when the plates that constitute the Earth's surface (see CONTINENTAL DRIFT) move over them.

W

Water cycle refers to the continual vaporization and condensation of the Earth's water. The oceans and land masses release water vapour as a result of solar heating. This vapour rises high in the atmosphere and is carried by surface winds until it descends once again as precipitation falls on the land and makes its way either directly or indirectly to the sea: it seeps into the surface to join with the ground water, or drains into rivers and streams. The oceans, because they contain over 97 per cent of the Earth's total supply of water, play an important part in this cycle – see also HYDROSPHERE.
Wave is a disturbance that moves through or over the surface of the ocean (or the land). The speed of a wave depends upon the properties of the medium. The ocean SWELL advances at an average rate of about 56 kilometres (35 miles) per hour in the Pacific and slightly less in the Atlantic.
Wave energy is the capacity of waves to do work. Typically, the ocean SWELL arriving at the shore releases something like 40 kilowatts of energy (about the power of a small family car) per metre of coastline. Research is underway to develop devices able to capture wave energy before it is dissipated as breaking waves on the shore.
Whale is the common name given to the larger members of a group of mainly marine mammals known scientifically as Cetacea. Most of the smaller members are called dolphins or porpoises. The Cetacea can be divided into three major kinds: an extinct group of toothed whales; the modern toothed whales; and the baleen or whalebone whales. The baleen whales are unusual in that their principal source of food is the PLANKTON, particularly KRILL. The toothed whales feed on fish, squid and even seals and penguins.

Z

Zonation of the seashore is possible according to the types and species of animal and plant that live on it. For example, the distribution of winkles, a small marine snail, varies according to species on the same shore. The same can be seen in the various BROWN ALGAE found on rocky shores.
Zooxanthellae are unicellular ALGAE that live in the tissues of marine creatures such as CORAL and the giant clam *Tridacna*. The clam literally 'farms' the algae as a supplementary source of food. The zooxanthellae live in the tissue that lines the shell – the mantle. The mantle is extruded from the shell and bent back over it. The algae thus receive the necessary sunlight for photosynthesis. In the course of time, some of the zooxanthellae are engulfed and digested by the body cells of the clam, providing it with an additional supply of food.
Zooplankton – see PLANKTON

A blue whale.

Index

Note: Page numbers in *italics* refer to illustrations; page numbers in **bold type** refer to entries in the glossary, between pages 84 and 91.

A
Abyss *10*, 14, 40, *60*, **84**, *89*
ACMRR **84**
Agar 72, *72*, **84**
Alaska 28
Albatross *34*
Alexander the Great 50
Algae *34*, 72, *72–73*, **84**
Alternation of generations **84**, 90
Aluminaut 54
Alvin 54
Amoco Cadiz 80
Amphipods **84**
Anadromous fish **84**
Angel fish *42*
Angler fish *34*, *35*, 40, *40*
Animal groups 36
Antarctic ice pack 30, *30*, *31*
Antarctic Ocean 10, 41
Aphotic zone **84**
Aphrodite aculeata see Sea mouse
Aqaba, Gulf of *12*
Aquaculture **84**, *84*
Aqualung 49, *65*
Archaeological excavation *64–65*, 91
Arctic Ocean 10, *10*, 30, 41, 82
Armoured fish *36*, 37
Astronesthes 40
Atlantic Ocean 10, *10*
Atoll **84**
Azoic **84**

B
Baleen whale *35*, *39*, 91
Baltic Sea 27
Barcelona *76*
Barrage 27, *28*
Barrel **84**
Barrier reef **84**
Bathypelagic zone *35*, *89*
Bathyscaphe 16, *16*, **84**
Bathysonde *17*
Bathythermograph 16, *17*
Bay *69*
Beach 29, *67*, 68, *71*, **84**
Beadlet sea anemone *71*
Beaufort wind scale **84**
Benguela Current *22*, 23
Benthic division **84**, *89*
Benthos **84**
Bering Land Bridge 28
Bering Strait 82
Bert, Paul 49
Bio-accumulation 59, **84**
Biogenous sediment **84**
Bioluminescence **84**
Biomass **84**
Bivalves **84**
Bladderwrack 72, *72*
Blowhole *69*
Bloom **84**, *86*
Blow torch 52, *52*
Blow-out *63*, **84**
Blue whale *39*, *39*, 91
Bonito *34*
Bony fish *36*, *37*, **84**
Bore **84**
Borisov, Pyotr 82
Boundary currents **84**
Brackish water **84**
Breakers **84**, *84*
Breathing mixtures *48–49*, 49
Brittle star *34*
Bromine 58, 72
Brown algae 72, *72*, **84**
Buoy *16*, *17*, *20*, **84**
Burrowers 70, *71*

C
Cables 52, *52*, 54
Calcareous ooze *15*
Camera *16*, 17, *46*, 52, *55*
Canaries Current *22*, 23
Caravel *84*
Cargo submarine *83*
Carrack **84**, *84*
Cartilaginous fish 36, **85**
Castellated iceberg 30, *31*
Catadromous fish **85**
Catspaws 24
Cave 68, *69*
Cay 74
Cephalopods **85**
Chaetognaths 36, **85**
Challenger, HMS 16, *16*, *89*
Channelled wrack *72*
Chemical elements 57, 58, *60–61*
Chemical pollution *80*, *81*
Chimaeras (rat-tail fish) *37*, *40*, 85
Christmas tree **85**
Cilia 36, **85**
Cliff 68, *69*, 70, *70*
Clownfish 42, *43*
Coal *60*, *60*
Coast *see* Shore
Coastal currents *see* Longshore drift
Cod *34*, 41, 45
Coelenterates 36, 84, **85**
Coin *64*
Columbus' ship *84*
Commensalism **85**
Compass *21*
Conshelf 51, *91*
Consub 55, 90
Container ship *20*, *20*
Continental crust *12*, 13, *13*, *60*, **85**
Continental drift *12*, 13, **85**, *89*
Continental rise *see* Continental shelf
Continental shelf *10*, 14, *14*, *60*, 62, **85**, *86*, *89*
Continental slope 14, *14*, *60*, *89*
Contouring raft 26, *26*
Convection currents **85**
Convergences **85**
Copepods *35*, **85**
Corals 36, *42*, *46*, **85**, *85*
Corer *17*, *17*
Coriolis, Gaspard de 23
Coriolis effect 23, **85**
Cousteau, Jacques-Yves 49, *51*, 54, *91*
Crab *36*, *37*, 70, *71*, 85, 91
Crust, Earth's *see* Continental crust
Crust, oceanic *13*, *60*
Crustaceans *34*, 36, 40, **85**
Cryopelagic fish 41
Current detector *17*
Currents *22*, 23, **85**
CURV *54*

D
Dahlia sea anemone *71*
Decca **85**
DDC (deck decompression chamber) *48–50*, 51, 90
Decompression *48*, **85**, *85*, 90
Deep-sea mining 57, 58, 60, *60–61*, **85**
Deep-sea zone *see* Abyss
Deep-sea creatures 40, *40–41*
Demersal fish *37*, *45*, **85**
Desalination 58, **85**, *85*, 86, 90
Destroyer *21*
Detritus **85**
Diamond *60*, *60–61*
Diatom ooze *15*
Diatoms **85**
Dictyoptoris membranacea 72
Dinoflagellates **85**, *89*

Distillation 58, **86**
Divergence **86**
Diving *46*, 47, 48, *48–49*, 90; and underwater homes 50, *50–51*; treasure exploration *65*, *64–65*; working conditions 52, *52–54*
Diving bell 50, *50*
Diving saucer *see Soucoupe*
Diving suit *48*, 49, *48–49*, **86**
Doldrums **86**
Dolphin *35*, *39*, *39*
Dounreay power station *76*
Dover, Straits of *81*, *82*
Drebbel, Cornelis 54
Dredging: port *76*, 82; seabed 17, *60*, 61
Drift ice 30
Drift net *see* Gill net
Drilling 62, **90**
Drilling rig *52*, *57*, 62
Drillship *62*, *63*
Dugong *39*
Dune *68*
Dyke 29
Dynamic positioning **86**
Dysphotic zone **86**

E
Earth: crust *12*, 13, *13*; proportion land and water *10*
Earthquake *12*, 13
Ebb tide **86**
Echinoderms *34*, 36, *42*, *71*, **86**, *86*
Echo-sounder 17, **86**, *86*
Ecosystem **86**
Eel 23, *35*
EEZ **86**
Effluent *80*
Electrical fishing **86**
Electrodialysis **86**
Emperor penguin *41*
Energy 26, 82, 83, *83*, **91**
English Channel *81*
Enteromorpha 72
Epeirogenic changes 28
Epipelagic zone *89*
Erosion 68, *69*
Euphausiid *41*, **87**
Euphotic zone **86**
Eustatic changes 28, **86**
Eutrophication *80*, **86**
Eutaemophorous 40

F
Factory ship *45*
Factory waste *80*, *80*
Falco, Albert 51
Fathom **86**
Fetch **86**
Filter-feeding **86**
Finch *75*
Fish *34*, *35*, 36; deep-sea 40; eggs 37; respiration 36
Fish farming *45*, *45*
Fish hook *71*
Fishing 44; bait *71*; electrical **86**; long-line **88**; net 70, **87**; over-fishing 45, **88**; trawl **86**, **91**
Fission 82
Fjord 29, **86**, *86*
Flatfish *34*
Flat wrack *72*
FLIP (Floating Instrument Platform) *89*
Float 23
Floe 30, **86**
Flood prevention 27, *28*, 29
Flying fish *34*
Food chain *see* Food web
Food pyramid *see* Food web
Food web 35, 36, **86**
Freefall sampler *16*
Freezing processes **86**
Fringing reef **86**
Fundy, Bay of 27
Fusion 82, *83*
Fyke net *70*

G
Galápagos Islands 75
Gametes 84, **86**
Gannet *34*, 70
Gaper *71*
Gas *60*, 62
Geneva Continental Shelf Convention (1958) **86**
Geophysical exploration **86**
GESAMP 81
Giant kelp *72*
Giant moa 75, *75*
Giant squid *35*, 85
Giant tortoise 75
Gill 36, *37*, **87**
Gill net *45*, **87**
Glacier 28, 30
Glass sponge *34*
Glomar Challenger 16, 17
Grab 17, *17*
Gravel *60*, *60*, **87**
Great Barrier Reef *42*, 74
Green turtle *74*
Green algae 72, *72*, **87**
Green turtle *34*
Grenadier *34*
Ground-effect principle *21*, **87**, *87*
Ground finch 75
Grouper fish *32*
Groyne 77
Gulf Stream *22*, 23, 82
Gulper eel *35*
Guyot 14, **87**
Gyres 23, **87**
Gyrocompass *21*

H
Haldane, John Scott 49
Halley, Edmund 50
Harbour *76*
Hawaiian Islands 74
Headland *69*
Heliox 51
Helium *48*
Hermit crab *37*, 70, 85
Heron Island 74
Herring *35*, 37
Highlands of Scotland *23*
High pressure nervous syndrome **87**
Holdfasts 72
Holiday resort 77, *77*
Holland (submarine) 54
Hovercraft *see* Ground-effect principle
Humboldt Current *see* Peru Current
Hydrocarbons 62, **87**, 90
Hydrofoil craft *20*, **87**, *87*
Hydrogen, heavy *see* Deuterium
Hydrography 16, **86**, **87**
Hydroids 35, *37*, 42, 70, **87**
Hydrophone 17, *17*, **87**
Hydrosphere **87**

I
Ice 30, *30*
Ice Age *28*, 28
Iceberg 30, *31*, 82, **87**
Icebreaker 31
Ice-cream 72
Ice patrol *see* International Ice Patrol
ICES **87**
Ice shelf *30*, **87**
Iguana 75, *75*
IMCO 81, **87**
Indian Ocean 10, *10*
Indicator species **87**
Industrial waste *80*, 81
Industry and the sea 76, *76*, 80, *80*
Inertial navigation system **87**, *87*
International Ice Patrol **87**
International law 82
Intertidal (littoral) zone **87**
IOC **87**
Iodine 72

Island *9*, 14, 28, 29, 74; man-made 82, *82–83*
Isobath **87**
Isostasy **87**

J
Jack-up platform 62, *62*, **87**
Jellyfish *34*, *36*, **87**
JIM (diving suit) 49, *49*

K
Kelp 72, *72*, **87**
Key *see* Cay
Killer whale *39*, *39*
Kiwi 75
Knot **87**
Krill *35*, 41, **87**, *87*

L
Labrador Current *22*, 23
Laminariales *72*, 87
Lamp shell 36, 40
Land reclamation 29
Lantern fish *35*, **87**
Laser profilometer 30
Lateen sail **88**
Laver bread 72, *72*
Lay barge **88**
Life, beginning of 33
Light penetration **88**
Light-producing fish 40, *40*
Limestone 68, 77
Limpet *34*, 70
Lisbon earthquake *24*
Little Mermaid, Copenhagen *38*
Lobster pot *71*
Lockheed oil production system *62–63*
Lockout submersible 51, *51*
Long-line fishing *44*, **88**
Longshore drift 68, **88**, *88*
Loran (Long Range Navigation) **88**
Lugworm *71*

M
Machinery under sea 52
Mackerel *34*, 37
Magellan **88**
Magnetometer *65*
Mammals 38, *38*, *39*
Manatee *38*, *39*
Manganese nodules *60*, *60–61*, 85, **88**
Mantle 13, *13*, *60*, **88**
Mapping *see* Hydrography
Marigram **88**
Malborough Sounds, N. Zealand 29
Marram grass *34*
Masked crab *71*
Maury, Matthew Fontaine 16
Maximum sustainable yield **88**
Mediterranean Sea 27, 81, 82
Mercury pollution 80
Mesopelagic zone *89*
Mid-Atlantic Ridge 13
Mid-oceanic ridges 13, **88**
Minamata Bay 80
Mineral-rich sands *60*, *60–61*, **88**
Mining: deep-sea 57, 58, 60, *60–61*, **85**; sands *76*
Mini-submarine 54
Missile 54, *54*, 90
Moa 75
Molluscs 36, 70, 84, 85, **88**
Monsoons **88**
Mudflat *69*, 70
Multistage flash distillation 58
Mussel *34*, 85

N
Nannoplankton **88**
Nansen bottle 16, *17*
Nautilus 54, 90
Navigational instruments *20*, 85, 87, **88**, *88*

Neap tide 27, *27*
Nekton **89**
Nemo 55
Neritic zone *89*
Net fishing *70*, **87**
Netherlands *29*
New Zealand coastline *29*
Nitrogen narcosis **89**
Nodules *see* Manganese nodules
North Atlantic Drift *22*, *23*, 23
North Equatorial Current *22*, 23
North Pacific Drift *22*, 23
North Pole 30, *30*
North Sea 28, 82, *83*
North Sea oil 63, *62–63*
Nuclear fission and fusion 82, *83*
Nuclear pollution 80, *81*
Nuclear submarine 54, *54*

O
Observation sphere *55*
Ocean: basin **89**; currents *22*, *23*; depth band *10*, *35*, *40*; exploration 16, *16*, *17*; floor 13, 14, *60*; pollution of 80, *80–81*; pressure *11*; surface *19*; temperature *10*; of tomorrow 79
Ocean, study of *see* Oceanography
Oceanic trench *10*, *11*, *12*, 13, 14, *35*, **89**, *89*
Oceanic zone *89*
Oceanography 16, **89**
Octopus *34*
Oil 57, *60*, 62, *62*, **90**; pollution *78*, 80, *80*, **89**; refinery *76*; rig *50*, *63*; storage tank *63*
Oil tanker 20, *21*, *62*, 80, *80*, *81*
Ooze *15*, **89**
Oscillating water column 26, *26*
OTEC (Ocean Thermal Energy Conversion) *23*
Oyashio *22*, 23
Oyster 59, *84*

P
Pacific Ocean *10*, *10*, 14; fisheries *44*, *45*
Pack ice *30*, **89**
Pagurus bernhardus see Hermit crab
Palm tree *23*
Pangaea *12*, **89**
Parrot fish *42*
Patrick Henry (submarine) **90**
Pearl *59*
Pebble beach *see* Shingle beach
Pelagic division **89**, *89*
Pelagic fish **89**
Pelagic sediments *15*, **89**
Penguin *41*, *41*
Peru Current *22*, 23
Pesticides 80, *80*
Petersen grab *17*
Phantom bottom **89**
Photic zone *see* Euphotic zone
Photosynthesis *35*
Phyla *36*
Phytoplankton 30, *30*, *34*, *36*, *41*
Piccard, Jacques 16
Pieces of eight *64*
Pilkington Bros 82, *83*
Pipeline *52*, *53*, *55*, *62–63*, 88
Pisces III 55
Plankton *35*, *35*, *36*, *41*, **89**
Plants *35*, *36*, 72, *72*, **84**, **87**
Plate *12*, 13
Platform *see* Jack-up platform; semi-submersible platform

Polaris missile *54*
Polar seas 41
Pollution *78*, *79*, *80*, *80–81*, **89**
Polychaetes **89**
Polyp 42, *85*, **89**
Porphyra 72
Porpoise 39
Port *76*, *76*, *83*
Portland stone 77
Port Royal, Jamaica 65
Portuguese man-of-war *87*
Power station *77*, *76–77*, *81*, *82*
Prawn *34*, *35*, 40
Precipitation *11*
Pressurized bases *see* Underwater habitats
Prevailing winds **89**
Probing tools *16–17*
Production platform *62*, 63, *63*
Project Atlantropa 82
Protozoa *36*, **89**
Purple heart urchin *71*
Purse-seine net *44*

Q
Quadrant *88*
Quarrying 77

R
Radioactive waste 80, *81*
Radiolarian ooze *15*
Radiolarians **89**
Ragworm 70, *71*
Raiatea Island *74*
Rance estuary *27*
Rat-tail fish *see* Chimaeras
Ray *34*
Razor shell *37*, *71*
RCV *55*
Red algae 72, *72*, **89**
Red clay *see* Pelagic sediments
Red Sea *11*, *12*
Red tide **89**
Reef *84*, *85*, *86*, **89**
Refraction of water waves **89**
Reproduction *90*
Reverse osmosis **90**
Rhodymenia 72
Ria *29*, **90**
Ridge *13*, *13*, **88**
Rotary drilling **90**
Russell rectifier *26*, *26*

S
Saline water *10*, **90**
Salt *10*, *58*, **86**
Salt dome **90**
Salter's ducks *26*
Salt pan *58*, *58*
Sand *60*, *60*; bar *68*; beach *67*, *68*, *68*, *69*, *69*, *70*; dune *68*
Santa Barbara oil leak *80*
Sargasso Sea *23*, *42*
Sargassum fish *40*, *42*
Sargassum weed *42*, *42*, *84*
Satellite, artificial *20*
Saturation diving *51*, **90**
Scallop *34*, *36*
Scarlet prawn *34*
SCOR **90**
Scuba diving *49*, *85*, *87*, *89*, **90**
SDC (submersible decompression chamber) *48*, *50*, *50*, *90*
Sea *see* Ocean
Sea anemone *34*, *36*, *37*, *37*, *42*, *43*, *70*, *71*
Seabed, study of *see* Oceanography
Sea-bathing *77*
Seabug *53*
Sea City *82*, *83*
Sea cow *38*, *39*
Sea cucumber *34*, *35*, 40

Sea Flea 54
Seagull *34*
Sea ice 30, *30*, **90**
Seal *38*, *38*, 41, *41*
Sealab 51
Sea lemon *71*
Sea lettuce *72*
Sea level 28, *28–29*
Sea-lion *38*, *38*
Seamount 14
Sea otter *38*, *38*
Sea potato *71*
Seashore *see* Shore
Sea spider *34*
Sea squirt *36*, *42*
Sea urchin *36*, *71*
Seawall 77; of the future 82
Sea water, composition of 11; extraction of minerals from 58, 59
Seaweed 72, *72–73*, **84**, **87**; farming *72*, *73*
Sediments 14, *15*, **90**
Seismic profiling 17
Seismic sea wave *see* Tsunami
Semi-submersible platform 62, 63, **90**
Sewage *80*, 81, *81*
Sextant *21*, *88*
Shadow zone **90**
Shark *35*, 36, 37, *37*
Shell *60*
Shingle beach *68*, *69*, *69*
Shipping lanes *20*
Ships *20*, *21*; factory ship *45*; fishing fleet *44*, *44*, *45*; oceanic research *89*; and pollution *81*; sunken 65; traffic zone *81*
Shore *28*, *29*, 29, *67*, *67*, *68*, *68–69*; life on 70, *70–71*; uses of *76*, *76–77*
Siebe, Augustus *48*
Skate *37*, *85*
Sledge *61*
Sonar **90**; in Arctic *30*; in oceanography 16, 17
Sonobuoy **90**
Sörgel, Herman 82
Soucoupe 54
Sounding **90**
Sound Navigation And Ranging *see* Sonar
South Pole *30*, *30*
Sperm whale *35*, *39*
Sponges *34*, *36*, *36*, *37*
Sporophyte see Alternation of generations
Spring tide *27*, *27*
Squid *34*, *35*, *36*, *36*, 40, *84*, *85*
Starfish *see* Echinoderm Station **90**
Storm surge **90**
Strait **90**
Subduction zone *see* Oceanic trench
Submarine *54*, *54*, *83*, **90**, *90*
Submersible *51*, *51*, *52*, *54*, *55*, **90**
Subsea completion **90**
Suez, Gulf of *12*
Sugar kelp *72*
Sundaland *28*
Sunlight zone *35*
Supertanker *20*, *21*, 76
Surtsey Island 74
Swash *68*, *92*
Swell *24*, **90**, *91*
Swim bladder *37*
Swordfish *35*, *37*
Symbiosis **90**

T
Tabular iceberg *30*, *31*
Tanker *see* Oil tanker
Tectonic changes 28
Tektite 50–51
Television camera *52*, *54*
Tellin *71*

Tern *35*
Terrigenous deposits *15*, **91**
Test drilling 63
Thallus **91**
Thames barrage *28*
Thermal power *23*, *83*
Thermocline *11*, **91**
Thongweed 72
Thorium 88
Tidal barrage *27*, *28*
Tidal power 27
Tidal wave *see* Tsunami
Tides *27*, **91**
Titanic 31, *87*
Tongue Point, N. Zealand *29*
Tools *18*, *52*, *61*, *64–65*
Toothed whale *39*, **91**
Toredo see Ship worm
Torpedo *54*
Torrey Canyon 80
Tortoise *75*
Trade winds *23*, **91**
Trawl *44*, *86*, **91**
Trawler *44*, *45*
Treasure excavation *64–65*, *65*
Trench *see* Oceanic Trench
Trident, USS *54*
Trieste 16, *16*
Trimaran *21*
Tripod fish *35*
Trophic level **91**
Tropical seas 42
Tsunami 24, *24*, **91**
Tubeworm *36*, 42, 70
Tuna *35*, *37*, *45*
Tunicates **91**

Turbidity currents **91**
Turtle *34*, *74*
Twilight zone *35*, *40*

U
Underwater habitats *51*, **91**, *91*
Upwelling *22*, *23*, *44*, **91**

V
Vacuum cleaner, seabed *64*
Vapour compression **91**
Verne, Jules *50*
Volcanic centre **91**
Volcano *8*, *12*, 13, 14

W
Walrus *38*, *38*
Walsh, Donald 16
Warbler finch *75*
Warship *21*, *21*
Water, manufactured 31, *58*, *58*, *82*, *85*
Water cycle *10*, *11*, **91**
Wave *18*, *24*, *68*, **89**, **91**
Wave energy *26*, *83*, **91**
Wegener, Alfred 13
Wesley, Claude 51
Whale *35*, *36*, 39, *39*, **91**, *91*
White-tip shark *35*, *37*
Wind *23*, *89*, *91*
Woodpecker finch *75*
Worm *36*, *37*, 40, *42*, *70*, *89*

Z
Zonation **91**
Zooxanthellae *42*, **91**
Zooplankton *35*, *35*

ACKNOWLEDGEMENTS
Picture Research: Jackie Cookson

Photographs: page 4/5, 6 Seaphot; 7 British Tourist Authority *left*, Seaphot *centre top, centre bottom, top right*, Pilkington Brothers *bottom right*; 8/9 ZEFA/Photri; 12 NASA; 16 Scripps Institution of Oceanography *centre*, Popperphoto *bottom*; 16/17 Mansell Collection; 18/19 Seaphot; 22/23 British Tourist Authority; 23 Heather Angel; 24 Ann Ronan Picture Library; 25, 27, ZEFA; 28 ZEFA *top*, Greater London Council *bottom*; 29 G. R. Roberts *top, centre*, ZEFA *bottom*; 31 Bruce Coleman; 32/33 ZEFA; 34, 36 Seaphot; 37 ZEFA *left*, Heather Angel *top right, centre right*, Seaphot *bottom*; 38 Bruce Coleman; 38/39 Pat Morris; 39 Heather Angel; 40, 41, 42, 43 Seaphot; 44 ZEFA; 45 White Fish Authority *top*, Novosti *bottom*; 46/47 Seaphot; 48/49 Shell; 50 Mansell Collection; 50/51, 52 Seaphot; 53 Seaphot, UDI Operations *inset*; 54/55 Seaphot; 55 Vickers Oceanics *top*, Seaphot *bottom left*, Hydro Products *bottom right*; 56/57 British Petroleum; 58 ZEFA, 59 Australian Pearl Co *top*, Wasson *bottom*; 61 HAM; 63 Shell, 64, 65 Seaphot; 66/67, 69 ZEFA; 70 Bruce Coleman; 70/71 ZEFA; 71 Seaphot; 73 ZEFA; 74 ZEFA *top*, Bruce Coleman *centre*, Seaphot *bottom*; 74/75 Seaphot; 75 Bruce Coleman; 76 Australian News and Information Bureau *top*, ZEFA *centre, bottom*, 76/77 UKAEA; 77 French Government Tourist Office; 78/79 Daily Telegraph Colour Library; 82 Marex; 82/83 Pilkington Brothers Ltd; 83 UKAEA; 84 ZEFA *left*, Mansell Collection *right*; 85 Seaphot *top*, Alan Hutchinson *centre*, Biofotos *bottom*; 86 Heather Angel *left*, ZEFA *right*; 87 Marconi Avionics *top*, British Tourist Authority *bottom left*, Seaphot *right*; 88 Mansell Collection; 89 Scripps Institution of Oceanography *left*, Bruce Coleman *right*; 90 US Navy; 91 Cousteau Society.

Endpapers: Seaphot; **Cover:** *Front* Seaphot; *Back* Seaphot *left, centre*, ZEFA *right*.

Artwork: Tom Brittain 20/21, 44 *centre*, 51, 55; ARTBAG Pavel Kostal 14/15, 23, 30 *top*, 48/49 *bottom*, 50, 68/69; LINDEN ART Graham Allen 34/35 *bottom*, 44 *bottom*, 70, 71 *centre, bottom*, 72 *top*, 75, Jim Channel 36, 38, 39; TUDOR ART AGENCY Norman Cumming, 12, 12/13 *bottom*, 15 *inset*, 20 *bottom left*, 22 *top, bottom*, 28, 30 *left*, 34/35 *top*, 60/61 *top*, 72 *bottom*, Ron Jobson 62/63, 80/81, Brian Pearce 48 *left*, 49 *bottom*, 62 *top*, Bernard Robinson 41, Mike Saunders 10, 11, 12/13 *centre*, 13 *right*, 16/17, 22 *left*, 24, 26, 27, 29, 42, 49 *top*, 54, 58, 60 *top, bottom*, 61, 62 *bottom*, 71 *top*, 81 *top*. **Cover design:** H.P. Lim.

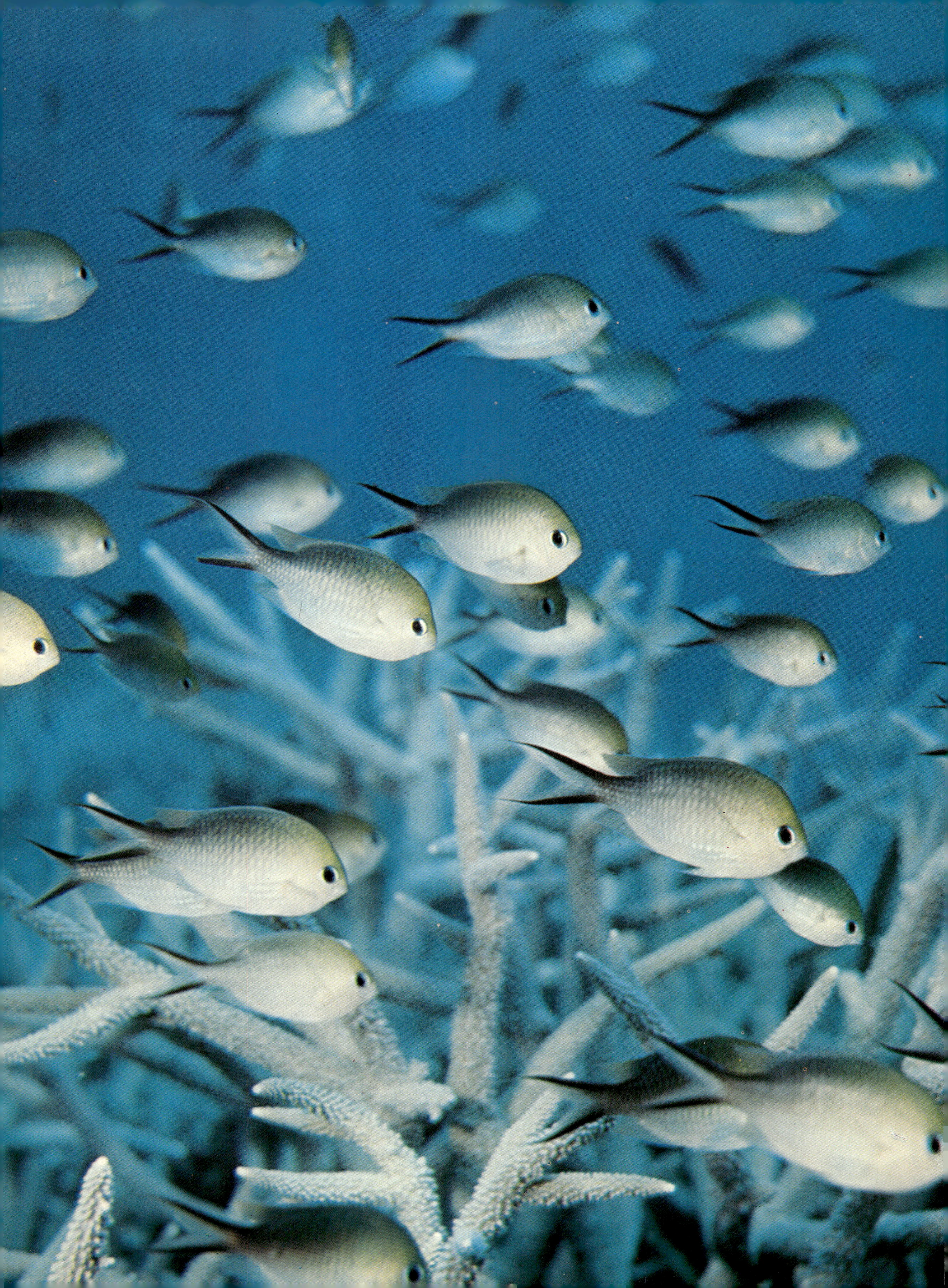

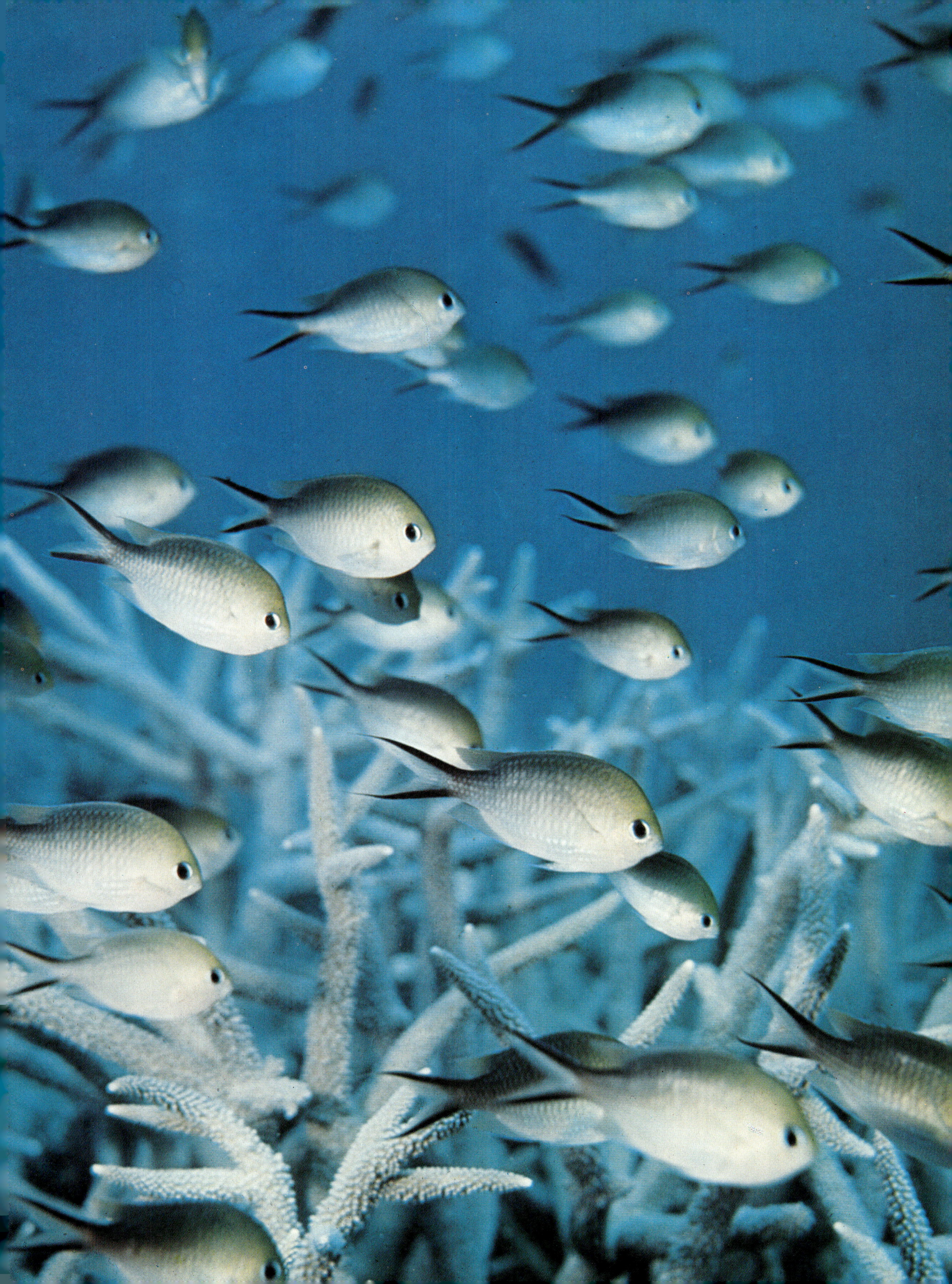

RETRO